Toward Affordable Systems II

Portfolio Management for Army Science and Technology Programs Under Uncertainties

Brian G. Chow, Richard Silberglitt, Scott Hiromoto,
Caroline Reilly, Christina Panis

Prepared for the United States Army
Approved for public release; distribution unlimited

ARROYO CENTER

The research described in this report was sponsored by the United States Army under Contract No. W74V8H-06-C-0001.

Library of Congress Control Number: 2011924894

ISBN: 978-0-8330-5126-4

Published 2011 by the RAND Corporation
1776 Main Street, P.O. Box 2138, Santa Monica, CA 90407-2138
1200 South Hayes Street, Arlington, VA 22202-5050
4570 Fifth Avenue, Suite 600, Pittsburgh, PA 15213-2665
RAND URL: http://www.rand.org/
To order RAND documents or to obtain additional information, contact
Distribution Services: Telephone: (310) 451-7002;
Fax: (310) 451-6915; Email: order@rand.org

Preface

Applying and expanding RAND's portfolio analysis and management (PortMan) method, this monograph seeks to help the U.S. Army select and manage its Science and Technology (S&T) programs to develop effective and affordable systems. Specifically, the results presented here expand the method and model described in this monograph's companion, *Toward Affordable Systems: Portfolio Analysis and Management for Army Science and Technology Programs* (Brian G. Chow, Richard Silberglitt, and Scott Hiromoto, MG-761-A, 2009), to include consideration of the actual capability gaps identified by the Army Training and Doctrine Command (TRADOC)/Army Capabilities Integration Center (ARCIC), in addition to consideration of uncertainty in the success of S&T programs. However, it should be emphasized that this monograph, like its companion, focuses on methodology development. Although we do use 2007 data from real Army S&T projects, we do so only to provide a more realistic demonstration of the methodology and its applications. Since resource limitations kept us from estimating all input parameters accurately enough to inform decisions about actual S&T projects, readers should not draw conclusions about the merits or drawbacks of any specific projects that we used for demonstration purposes.

This monograph should be of interest to S&T and acquisition managers who are responsible for portfolio management of programs; engineers in research, development, test, and evaluation programs; and those who are interested in the optimal allocation of funds among different programs and/or developmental stages to yield the lowest total lifecycle cost in meeting future capability gaps. The model feature devoted to handling uncertainty may be of particular interest to many others.

This research was sponsored by Stephen Bagby, Deputy Assistant Secretary of the Army (Cost and Economic Analysis), Office of Assistant Secretary of the Army (Financial Management and Comptroller), and it was conducted within RAND Arroyo Center's Force Development and Technology Program. RAND Arroyo Center, part of the RAND Corporation, is a federally funded research and development center sponsored by the United States Army. For further information, contact the principal investigators, Richard Silberglitt (email Richard_Silberglitt@rand.org, phone 703-413-1100 extension 5441) or Brian Chow (email Brian_Chow@rand.org, phone 310-393-0411 extension 6719).

The Project Unique Identification Code (PUIC) for the project that produced this document is SAFMR08810.

For more information on RAND Arroyo Center, contact the Director of Operations (telephone 310-393-0411, extension 6419; FAX 310-451-6952; email Marcy_Agmon@rand.org), or visit Arroyo's website at http://www.rand.org/ard/

Contents

Figures

Tables

Summary

S&T plays a central role in the ability of the Department of Defense (DoD) to field the advanced weapons systems that give the U.S. military its unmatched technological superiority. In view of this fact, DoD's program for acquiring new systems has long been linked with its S&T programs in basic and applied research and advanced technology development.[1] In 2008, for the first time, DoD explicitly documented a process for technologies coming out of its S&T programs to be integrated into new systems at every stage of acquisition (see Figure S.1). Under previous policies, the linkage between S&T and acquisition typically occurred at the initial stage of the acquisition process, prior to Milestone A, while the basic system concept was being refined and before the total costs to develop, field, and operate a new system were assessed. These lifecycle costs were part of an analysis of alternatives (AoA) due at Milestone A in the acquisition process (shown in Figure S.1), just before a new system entered the "technology development" stage. Consequently, the military's S&T planners were not called upon to consider the lifecycle costs of the systems in which new technologies were being used until an AoA was required. They had the costs of their own S&T programs to manage but could do so more or less independently of total systems costs. The 2008 policy reflects a department-wide emphasis on technology insertion at every stage of the acquisition process. To support such insertions, planners would need to make lifecycle cost estimates at the completion of the S&T programs, since the next step, the AoA analysis, where lifecycle cost estimates are typically made, may be skipped.

A second significant change in the acquisition policy environment occurred in 2009 when Congress established a DoD position, the director of Independent Cost Assessment. The director is tasked with conducting independent estimates of the cost of new major military systems. For any new system, this director needs to consider "tradeoffs among cost, schedule, and performance objectives" (P.L. 111-23, p. S. 454-1).

1 Advanced technology development "includes development of subsystems and components and efforts to integrate subsystems and components into system prototypes for field experiments and/or tests in a simulated environment." See Office of the Under Secretary of Defense, Comptroller, *DoD Financial Management Regulation, Volume 2B: Budget Formulation and Presentation*, "Research, Development and Evaluation Appropriations," Chapter 5 (July 2008), July 6, 2000, p. 5-2.

Figure S.1
Links Between Army S&T Programs and the Defense Acquisition Management System

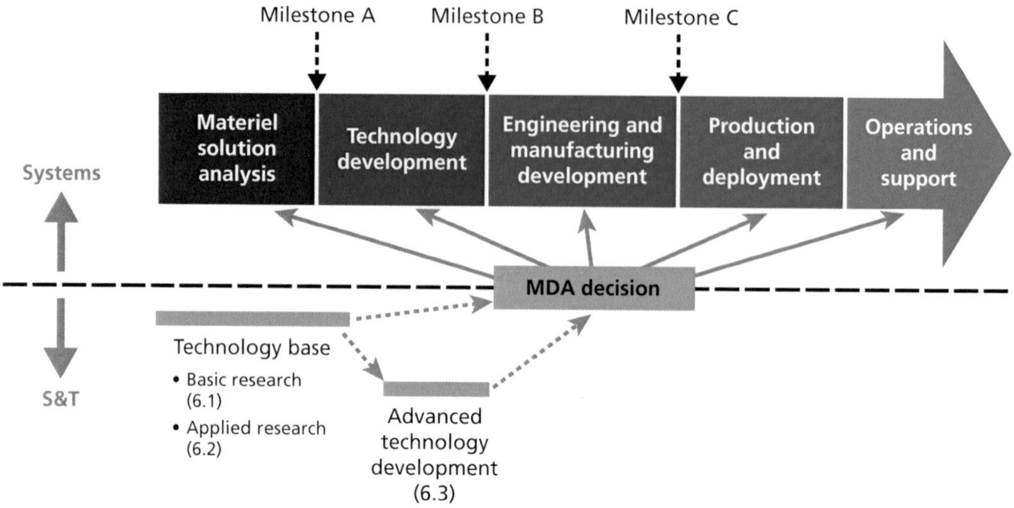

SOURCE: Simplified graph from Brown, 2008.
RAND *MG979-S.1*

Both of these policy developments highlight a rising concern with the cost of advanced weapons systems. Therefore the Army, as well as the other services, will need to rely more than ever on its scientists and engineers to design systems that will not only provide necessary capabilities, but that can also be acquired and operated affordably. Decisions made in the S&T stages of the acquisition process are vital in determining the total lifecycle costs of a new system. Historically, 70 percent of the overall cost of a system is incurred once it enters the final Operations and Support phase (U.S. Army, 2006, pp. 26–27). It is much easier to take steps to reduce those late-stage costs early in the acquisition process, before a design is fully developed and hardened. In this way, performance and lifecycle costs can be weighed against each other to determine what trade-offs may be viable.

The Army faces continual challenges in selecting portfolios of S&T projects that will meet future capability needs at an affordable overall cost. Uncertainty about whether all funded S&T projects will succeed compounds these challenges, as does the ever-present possibility that changes in the economic or strategic environment will compel some capability requirements to be altered.

A Process to Make Early-Stage Decisions That Will Lead to Effective and Affordable Systems

With this need in mind, the RAND Corporation developed a process to help the Army incorporate lifecycle cost into S&T planning under uncertainty.[2] This process offers several advantages. It equips the Army to consider lifecycle costs at the S&T stage so that the Army can design more cost-effective systems. It can complement the Army's existing process for managing its S&T programs. Moreover, the process will create a new opportunity for dialogue among stakeholders and allow different viewpoints and perspectives to be analyzed objectively in the process of building an S&T portfolio.

The Army's S&T managers can use the RAND process as an aid in carrying out two fundamental planning tasks:

- **Mapping supply and demand:** The process first broadly identifies where the Army may encounter problems meeting requirements (the demand) with existing S&T projects (the supply). It then refines this broad map of supply and demand, taking into account that some S&T projects in the portfolio will not lead to a fielded system, because either the project fails or the system is not as cost-effective as another.
- **Selecting an optimal S&T portfolio:** The process next identifies how the Army's S&T planners can manage expected or unexpected budget cuts, while keeping chances of meeting requirements as high as possible.

Mapping Supply and Demand

With any S&T portfolio, inevitably certain projects will succeed, while others will fail. (For the purposes of this study, we consider a successful S&T project one that meets its system performance and cost objectives and is suitable for further development into a fielded weapons system.) Failure to take this reality into account during planning can create a false impression of how well a given S&T portfolio will meet the Army's capability needs. An accurate map of mismatches in supply and demand is the product of two screening steps, one of which accounts for uncertainty. In the first step, the RAND process helps planners establish a baseline for measuring how well supply meets demand by assuming that all S&T projects in a given portfolio will succeed. But, potentially, enough projects could fail so that all requirements cannot be met. Consequently, in a second step, the process enables planners to analyze how this uncertainty affects that portfolio's ability to meet requirements. Through this process,

[2] The process is presented in two monographs. The first—Chow, Silberglitt, and Hiromoto, 2009 (hereafter referred to as TAS-1)—provides the basic model. The current study (TAS-2) adds a simulation model to deal with uncertainty.

planners can clearly see which areas of demand are at risk of not being met and where the portfolio may need to be adjusted accordingly.

Step One: Broadly identify where the Army may encounter problems meeting requirements (demand) with existing S&T projects (supply). A comprehensive picture of supply and demand takes the form of a matrix (see Figure S.2). For the demand in our demonstration of the RAND process, we use the force operating capability (FOC)–gap requirements defined by 2006 data from TRADOC/ARCIC.[3] These requirements are displayed from left to right along the top of the matrix. For this demonstration, the Army Technology Objectives (ATOs) constitute the supply. We use these because they are the Army's highest-priority S&T projects and are examples of S&T projects that can lead to specific systems.[4] The ATOs are displayed from top to bottom along the left side of the matrix.[5] Each row in the matrix displays how many capability gaps a given ATO addresses within each FOC requirement. The columns indicate how many ATOs address that gap.

Drawing on this matrix and assuming that none of the ATOs in this portfolio will fail, we can identify the overall performance of the current set of ATOs and expose potential problem areas—that is, where capability gaps are at risk of not being met (see Figure S.3). Under these ideal conditions, where 100-percent success is a given, the overall set of ATOs in the pipeline can satisfy nearly all of the Army's current capability needs: Ten out of the 11 meet at least 100 percent of the FOC requirement. Only one problem area emerges clearly: Existing ATOs meet only 57 percent of requirement 10, "Training, Leadership, and Education."

However, more-detailed consideration of this map provides important information about where the Army's S&T planners might want to adjust the roster of ATOs in the existing portfolio. Two of the requirements—4 and 6—are only just met. Both constitute red flags, where the current set of ATOs is at risk of not being able to meet these needs should certain projects fail. These are areas where new ATOs may be needed. On the other hand, requirements 3, 5, and 8 are overmet, indicating redundancy among projects in the portfolio. Here may lie opportunities to reduce funding of some existing ATOs that are providing too high a level of redundancy. The RAND process can show how to save money by terminating some ATOs and using that money to fund some new ATOs targeted at requirements that will be inadequately met by existing ATOs, resulting in a higher chance of meeting all requirements at the same or even lower total cost.

---

[3] To allow for analysis according to FOC, RAND assigned each of the TRADOC/ARCIC–defined capability gaps to the FOC to which it best applied. In this monograph, the capability gaps are listed only by FOC and number. The Excel spreadsheet that identifies the gaps came from an unpublished TRADOC/ARCIC report on force-capability-gap analysis using 2006 data.

[4] Any desired set of S&T projects expected to lead to specific systems could be analyzed using the RAND process.

[5] The ATOs are listed by technology type and title.

Figure S.2
Matrix of Supply and Demand (Gap-Space Matrix)

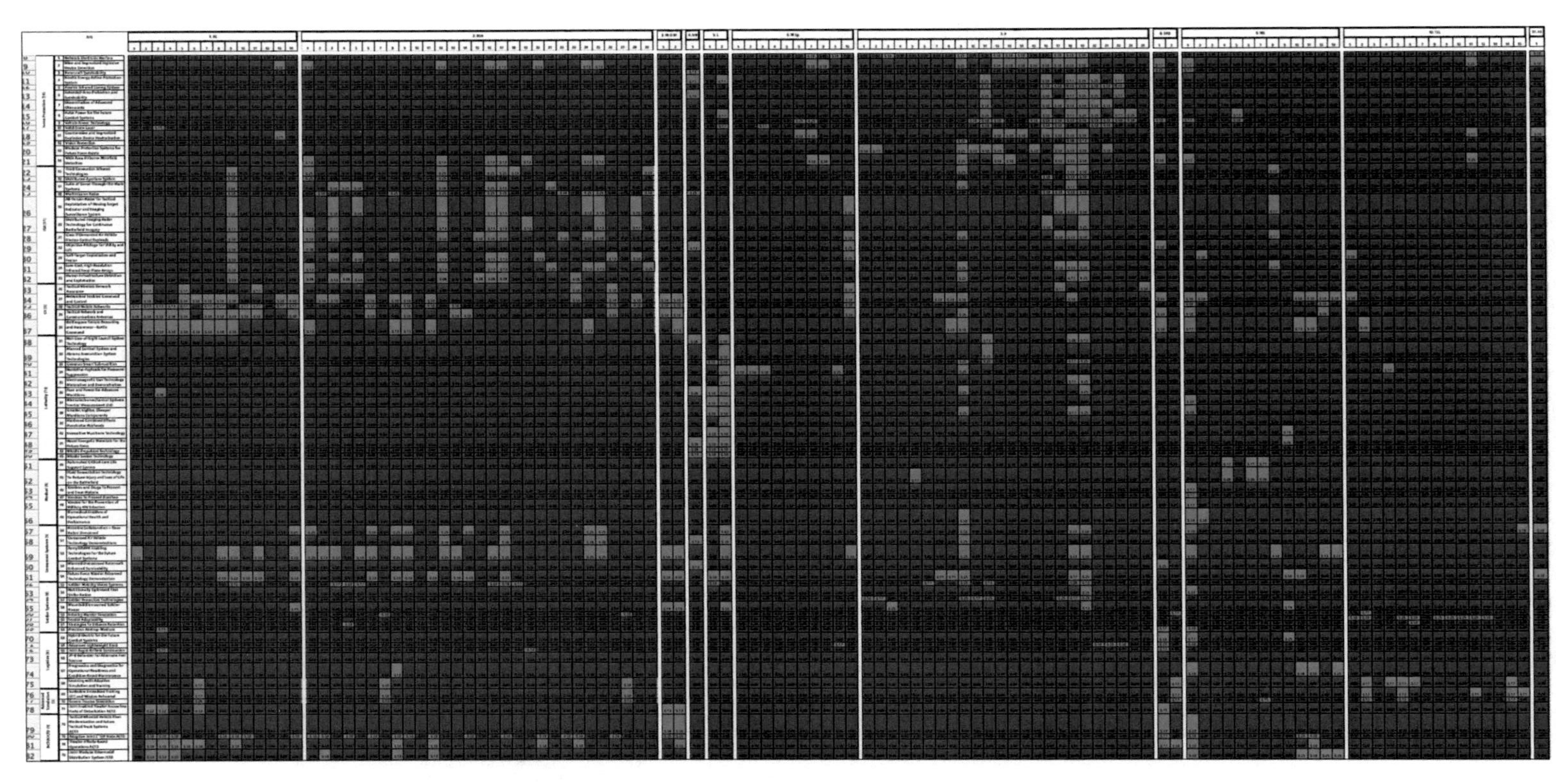

NOTE: A larger version of this matrix is shown in Chapter Three, Figures 3.1–3.8.
RAND *MG979-S.2*

Figure S.3
Map of the Match Between Supply and Demand, Assuming a 100-Percent Success Rate for All ATOs in the Existing Portfolio

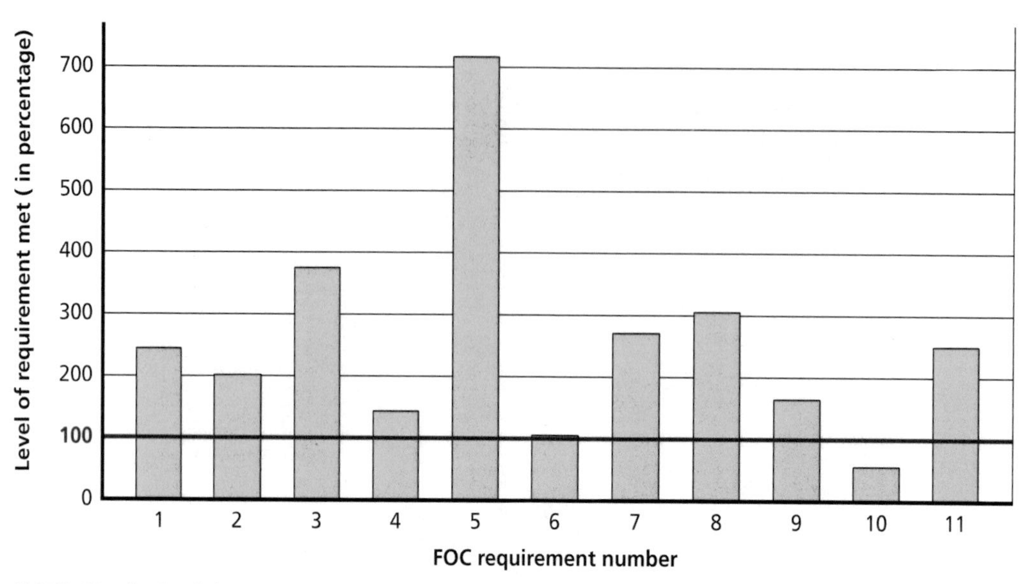

NOTE: We obtained these estimated percentages by adding the independent contributions of each ATO to each capability gap using the gap-space method defined in Chapter Two, as described in Chapter Three.
RAND *MG979-S.3*

This initial map of supply and demand, which assumes a 100-percent rate of success for all ATOs, is a solid starting point for planning efforts. It shows the Army's S&T managers the overall performance of existing ATOs and gives them a broad idea of where they might divert funds toward new ATOs that target unmet requirements or those at risk. The levels of each FOC requirement met—that is, the 11 percentages displayed from left to right in the bar graph—constitute a baseline that planners can use to refine the map in the next step of the process.

Step Two: Refine this map of supply and demand, given that some S&T projects in the portfolio will not lead to a fielded system. Assuming a 100-percent rate of success for all ATOs, although informative, is ultimately a theoretical exercise: Some S&T projects will inevitably fail. Because of this, the Army cannot be sure that existing projects will meet all of its capability gaps. When the success of S&T projects is uncertain, planners must unavoidably deal with probability. Our concept of a *feasible percentage* is a way of managing the S&T portfolio under uncertainty. The feasible percentage indicates the probability that a given portfolio of S&T projects will meet a defined requirement level for each FOC, with the possibility built in that some projects in the portfolio will fail—however, without knowing in advance which ones these will be.

The RAND process takes account of uncertainty by employing a model that uses as input the probability of success of each S&T project in the existing portfolio. To demonstrate this model, we run a computer simulation 10,000 times that follows three principal steps:

- Assume a 10-percent probability of failure for all ATOs.[6]
- Allow projects in the existing portfolio to fail randomly at that rate.
- Estimate the effects.

The feasible percentage is the percentage of times out of those 10,000 runs that the portfolio meets a defined requirement level for each capability gap.

In our demonstration of the RAND process, when ATOs in the existing portfolio fail randomly at a rate of 10 percent, as a set, they fall far short of meeting the Army's baseline percentage levels. Recall that the baseline is the outcome of Step One in the process of mapping supply and demand: namely, the percentage of each FOC requirement met by the current portfolio of ATOs *when all projects in that portfolio succeed.* Without the possibility of any ATOs failing, the overall match between supply and demand is very good; the existing portfolio can meet 100 percent of all the requirements except for 10, which is met at 57 percent. But with the 10-percent failure rate built in, the chance that the current portfolio will meet these baseline percentages—what we call the *feasible percentage*—drops sharply. Even with a 90-percent rate that any given project will succeed, the existing portfolio of ATOs will likely have only a 16-percent chance of matching the baseline FOC percentages (100 percent for each FOC requirement except 57 percent for requirement 10). At this rate of failure, the existing ATO portfolio would clearly not be meeting all of the percentages in the baseline. But it is unclear how much each percentage in the baseline needs to be reduced in order for the portfolio to have a good chance of meeting these lowered requirements.

Consequently, the Army needs a way to refine the initial map of supply and demand when the possibility of failure is introduced. The RAND process includes such a routine. Within each FOC requirement, a number of ATOs will contribute to it. For illustration, assume that these ATO contributions together amount to 110 percent of a specific FOC requirement. The routine starts by asking what if the ATO making the largest contribution, say 30 percent, failed to be completed. Then, the rest could at best meet 80 percent (i.e., 110 – 30 percent) of that FOC requirement. Critical to the routine is a different perspective: If that requirement were to be reduced to 80 percent, planners could afford a failure in any one of the ATOs that make a contribution to that requirement, since the 80-percent figure was arrived at assuming the *largest* contributor failed. Indeed, for the simulation (with the largest contributor to each FOC assumed to fail) with reduced requirements (to 88 percent in FOC 6, to 36 percent in FOC 10, but no reduction in others) shown in row two of Table S.1, the feasible percentage jumps

[6] The model allows different probabilities of failure for different projects. However, because planners generally do not know which projects will actually fail, using the historic failure rate for every project is a reasonable approximation. For this demonstration, we assume a single 10-percent failure rate.

from 16 to 73. The last row in the table shows that the feasible percentage increases to 99.8 if each FOC requirement is reduced to a level that can accommodate the failure of the four largest contributing ATOs. In other words, a simultaneous failure of any four ATOs within each category would still allow the set of reduced requirements to be met. The price to pay to drastically increase the feasible percentage from 16 to 99.8 is the need to reduce the requirements from 100 to 75 percent, 58 and 50 percent for FOC 4, 6, and 11, respectively; and FOC 10 from 57 percent down to 23 percent. This is an attractive price because we can design new ATOs targeted toward these requirement gaps to bring the requirements back to the baseline level (row one). The cost of doing so is much lower than that of adding new ATOs with a targeting pattern similar to the existing ATOs in order to bring the feasible percentage from 16 to 99.8.

Selecting an Optimal S&T Portfolio

Once the RAND process has been used to precisely determine which parts of the supply the existing ATOs should meet, with the rest to be met by new ATOs, it can then help the Army select an optimal S&T portfolio among the existing ATOs. In other words, only certain existing ATOs should be continued, and the money saved from the termination of the rest is more cost-effectively spent on new ATOs. The process also provides S&T planners a means of identifying how to manage (expected or) unexpected budget cuts, while keeping the chances of meeting requirements as high as possible. Budget cuts may be made to the total budget for developing the new systems and getting them through procurement, then fielding, operating, maintaining, and, finally, decommissioning them—the total implementation budget—or to the budget for funding ATOs to the end of the S&T phases—the total remaining S&T budget. Changes to either budget will alter the probability that ATOs will cover requirements—in other words, the feasible percentage. The sum of these two budgets is called the total remaining lifecycle budget.

The map of supply and demand deals purely with the ability of a portfolio to meet requirements. But in the real world of acquisitions, cost is an ever present concern. When costs are factored in, the goal of S&T planners becomes the *sweet spot*

Table S.1
Refined Map of the Match Between Supply and Demand

FOC Requirement Number and Percentage											
1	2	3	4	5	6	7	8	9	10	11	Feasible Percentage
100	100	100	100	100	100	100	100	100	57	100	16
100	100	100	100	100	88	100	100	100	36	100	73
100	100	100	75	100	58	100	100	100	23	50	99.8

NOTE: The final row is the new baseline.

between the probability of meeting requirements and the affordability of doing so. Such a sweet spot indicates the total remaining S&T budget and the total remaining lifecycle budget that should be spent on the selected existing projects and their systems and the feasible percentage that will result. If a planner wants a higher feasible percentage, the sweet spot tells us that it is less costly to fund new projects, as opposed to funding more existing projects. To find this sweet spot for existing projects, the S&T planner must make trade-offs between the feasible percentage and affordability. Our process shows us these trade-offs.

To explore what happens to the feasible percentage when budgets are reduced, the RAND process uses a linear programming model and simulation, as described in Chapters Two and Four. In our demonstration, we first adjust the baseline percentage levels for each FOC requirement to match the final map of supply and demand (i.e., the bottom row of Table S.1) and then assume a total remaining S&T budget of $3.1 billion, enough to allow funding of all the ATOs to completion of S&T (see Figure S.4). Under these conditions, as the figure shows, cutting the total remaining lifecycle budget nearly in half—from $138 billion[7] to $67 billion—gives the seemingly surprising result that the large cut has practically no effect on the feasible percentage: The

Figure S.4
The Effect of Cuts in the Total Lifecycle Budget on the Probability That Existing ATOs Will Meet Requirements

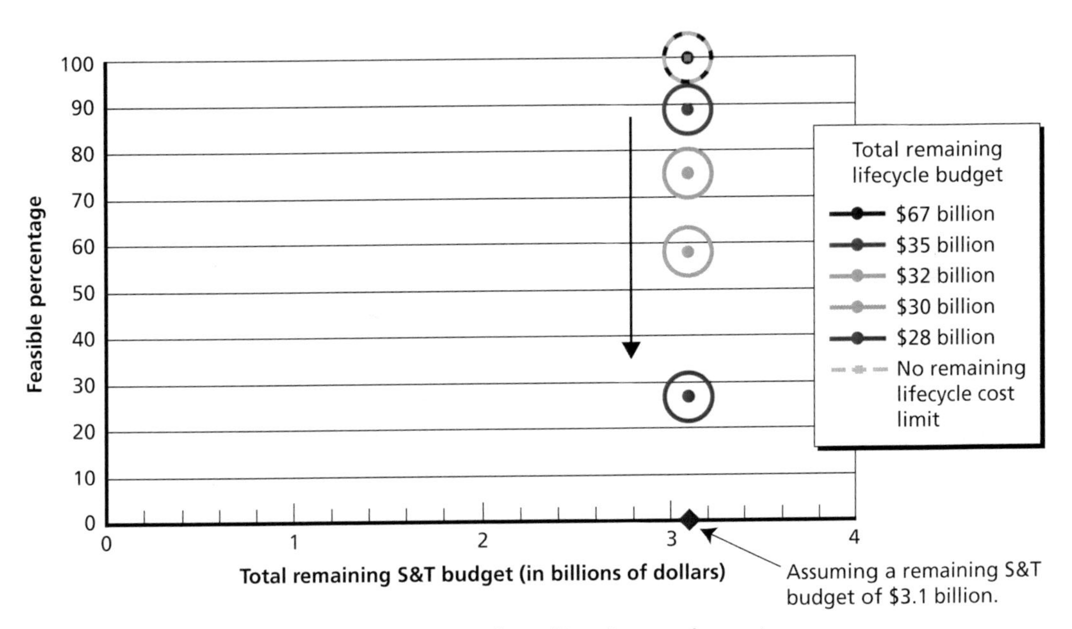

NOTE: The feasible percentage is based on an adjusted baseline requirement.
RAND *MG979-S.4*

7 This would be the total lifecycle cost if all systems resulting from the 75 existing ATOs were developed and fielded, designated "no total remaining lifecycle cost limit."

probability that existing ATOs will meet the requirements remains the same.[8] Cutting the lifecycle budget in half yet again only brings a small loss in feasible percentage. When the budget drops from $67 billion to $35 billion, the probability of meeting requirements becomes 88 percent. This is a big gain in affordability for only a small trade-off in probability. But the loss is not linear: Trimming only $5 billion more makes probability suddenly drop sizably, from 88 to 58 percent. Shaving off another $2 billion spurs a decline of 31 percent more. Given these trade-offs, the most cost-effective total lifecycle budget for the systems resulting from these ATOs is $35 billion.

Similarly, the S&T budget can be cut with some flexibility only up to a certain point (see Figure S.5). When we take the most cost-effective lifecycle budget of $35 billion as an anchor point and now focus on altering the remaining S&T budget, once that budget falls below $1.5 billion, the probability of meeting requirements plummets. As the figure shows, the same pattern holds within other total lifecycle budgets as well. This cautions the S&T planner not to drop the S&T budget below this $1.5 billion threshold. For the current case, with a lifecycle cost of $35 billion, the opti-

Figure S.5
The Sweet Spot Between Performance and Affordability in the Lifecycle and S&T Budgets

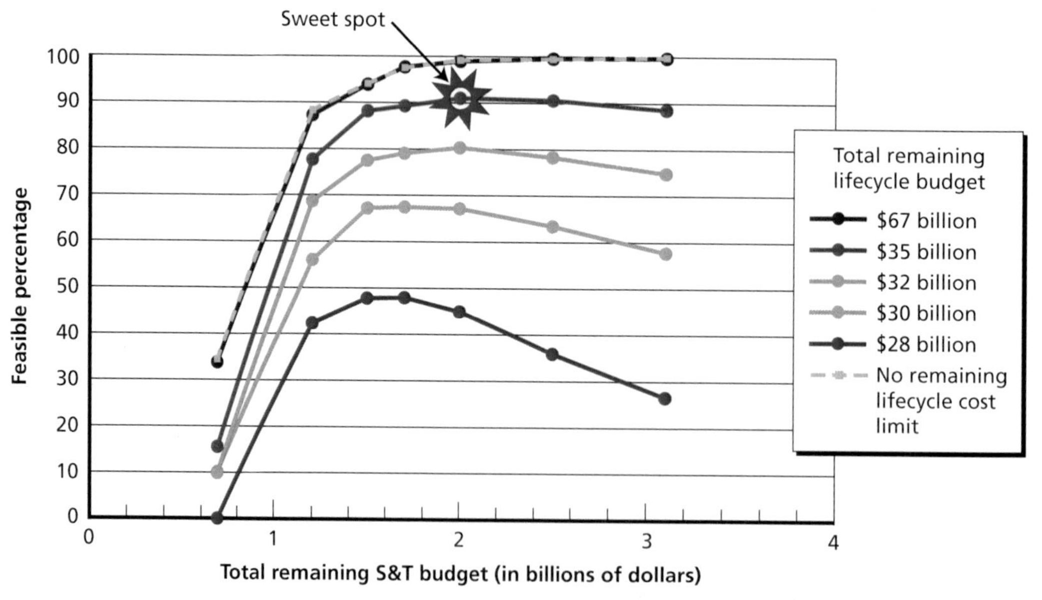

RAND *MG979-S.5*

[8] This result reflects a high level of redundancy in the systems resulting from the ATOs. Therefore, only some, and far from the complete set, of the successful ATOs are needed to be further developed into fielded systems in order to meet all requirements—and $67 billion will suffice.

mal S&T budget is $2 billion, because that amount allows the probability of meeting requirements—the feasible percentage—to reach its peak. Thus, the sweet spot for supporting the existing projects is a total remaining S&T budget of $2 billion and a total remaining lifecycle budget of $35 billion. There the Army will get a 91-percent probability (feasible percentage) of meeting its FOC requirements.

Once the lifecycle and S&T budgets are set at the sweet spot, the job then becomes to select the most cost-effective ATOs. Different portfolios of ATOs are of course possible for each combination of total and S&T budgets; various portfolios may perform better or worse. The RAND process includes an algorithm that automatically tries out different combinations of ATOs meeting any given set of budget constraints. Uncertainty is taken into account, with only successful ATOs included in the combinations. The results suggest which ATOs to keep and which to discontinue to get the highest possible feasible percentage—in other words, the most cost-effective ATOs.

To demonstrate, we take the optimal budgets from our analysis: $35 billion for lifecycle costs and $2 billion for S&T costs. Within these cost constraints, the model suggests 53 ATOs to keep and 22 to discontinue (see Figure S.6). In Figure S.6, we also compare the model selection with the best of many customary metrics we investigated. ATO 58 is rejected, even though its ratio of benefit-to-S&T cost is similar to many others selected. In contrast, three ATOs (48, 46, and 47) are selected although they have ratios even worse that those rejected ATOs with very poor ratios. These recommendations cannot be derived from a simple analysis of benefit-to-cost ratios; the model built into the RAND process is needed. It accurately represents the complex interplay between the costs of each ATO, requirements met by multiple ATOs, and the uncertainty in success of the ATOs.

Testing the RAND Process as Part of the Army's Annual Process

The RAND process could be integrated into the Army's existing annual process for making decisions about the S&T portfolio. This would be an effective way for the Army to test it, by using it to produce real-life results. The key stakeholders involved would be the S&T project managers, the Deputy Assistant Secretary of the Army for Research and Technology (DAS [R&T]), and the Warfighting Technical Council (WTC). It would take two fiscal years to run the whole test,[9] with the first year needed to establish a baseline. For the requirements, the Army could use the latest capability needs requirements provided by TRADOC/ARCIC.

[9] Of course, the test can be completed much sooner if the Army commits greater resources to it.

Figure S.6
The RAND Model's Selection of Those ATOs to Keep and Those to Discontinue

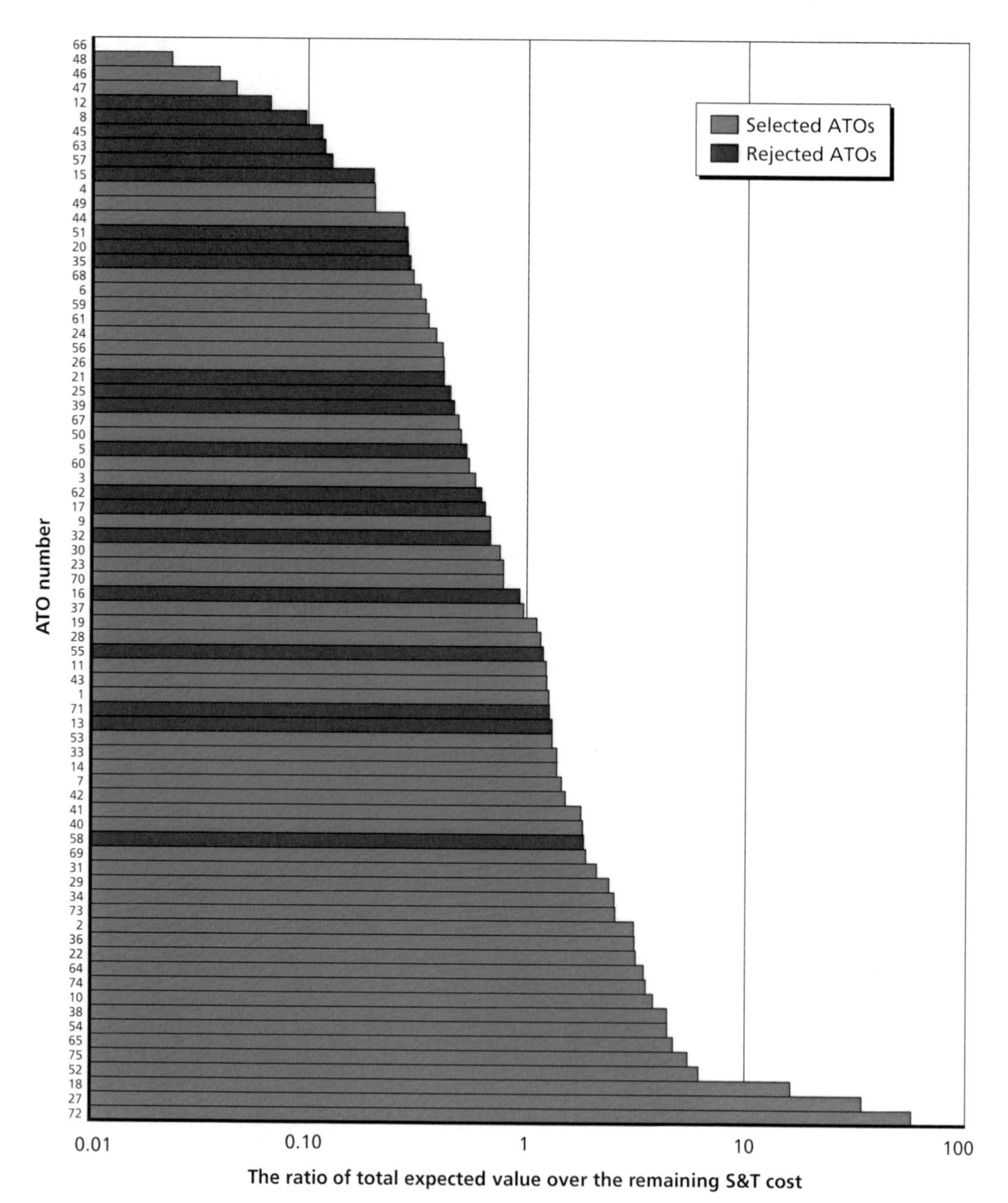

NOTE: The optimal budgets of $35 billion for lifecycle costs and $2 billion for S&T costs are used to derive the benefit-to-cost ratio.
RAND *MG979-S.6*

The First Year

In the first year, the S&T project managers and DAS (R&T) would work together to establish an initial baseline for both performance and cost. This process would follow several steps:

- S&T project managers would estimate (1) the contribution that each S&T project that leads to a fielded system will make to Army capabilities and (2) the lifecycle cost of the system derived from each S&T project.
- They would then provide these data to DAS (R&T).
- At the same time, DAS (R&T) would assemble data on the S&T cost to complete each project.
- Taking the estimates provided by the S&T project managers, DAS (R&T) would use a Delphi (or other) method to gather expert opinion on whether the estimates are appropriate, too high, or too low.
- DAS (R&T) would then apply RAND's process to inform S&T project managers of his office's current assessment of their projects. The process includes RAND's linear programming model and simulation (as described in Chapters Two and Four) and would provide outputs such as those shown in Table S.1 and Figures S.4–S.6.
- As the final step in the first fiscal year, DAS (R&T) would invite the S&T project managers to improve their baseline estimates.

The Second Year

In the second year, estimates would be refined and decisions implemented. WTC would now join the S&T managers and DAS (R&T) in the process.

- In the first six months of the year, the S&T project managers would provide their adjusted estimates and justify to WTC any differences from their baseline estimates.
- In the second six months of the year, DAS (R&T) would reapply RAND's process, with adjusted estimates where relevant, and provide the revised outputs to WTC.
- Toward the end of the year, WTC would make decisions based on the revised outputs and take any necessary corrective actions with respect to the S&T portfolio.

How Would Incorporating the RAND Process Improve the Army's Annual Process and Help It Better Manage Its S&T Portfolio?

RAND's process can help Army S&T planners monitor the expected performance of *and* the lifecycle cost of new Army weapons systems, and then weigh the trade-offs between the two, allowing adjustments to be made at the S&T stages. These early adjustments can improve the Army's ability to build affordable new systems that also satisfy the Army's capability requirements. Integrated into the Army's existing S&T decision process, the RAND process can be a useful management tool for Army S&T planners. It can enable them to make better decisions both about individual systems and about the Army's S&T portfolio as a whole. It can secure sizable cost savings in the long-term, making it less likely, for instance, that needed weapons systems will be cancelled because they are too costly. It can help planners deal wisely and effectively with suboptimal budgets. Finally, it provides a means with which they can manage the inevitable uncertainties involved in planning by permitting unbiased "what-if" analyses of variations in the expected performance and cost of a portfolio of ATOs when certain projects fail or budgets fluctuate, as in real life they are bound to do.

Acknowledgments

We acknowledge the advice, guidance, and financial support of David Henningsen of the U.S. Army Deputy Assistant Secretary of the Army for Cost and Economics, without which this study could not have been possible. We are pleased to acknowledge and thank our peer reviewers, Walter Perry and Lance Sherry, who provided many insightful comments and suggestions that contributed significantly to improving the exposition and explanation of our method, model, simulation, and results. Finally, we thank Bruce Held, our program director, for providing prompt and valuable comments during every stage of the project, from proposal to interim review to final report, and Susan Bohandy for her expert assistance in the development of the Summary.

Abbreviations

ACTD	Advanced Concept Technology Demonstration
AM	air maneuver
AoA	analysis of alternatives
ARCIC	Army Capabilities Integration Center
ATD	advanced technology demonstration
ATO	Army Technology Objective
BC	battle command
BLOS	beyond line of sight
BSA	battlespace awareness
BSTC	Budgeted Total Remaining S&T Cost
C4	command, control, communications, and computers
CPMR	capability portfolio manager
CPMT	capability portfolio management
DARPA	Defense Advanced Research Projects Agency
DAS (R&T)	Deputy Assistant Secretary of the Army for Research and Technology
DoD	Department of Defense
EMD	engineering and manufacturing development
EMDC	engineering and manufacturing development cost
EO	electro-optical
EV	expected value

FCS	Future Combat Systems
FOC	force operating capability
FY	fiscal year
HE	human engineering
HIV	human immunodeficiency virus
IED	improvised explosive device
IR	infrared
ISR	intelligence, surveillance, and reconnaissance
JCTD	joint capability technology demonstration
L	lethality
LPM	linear programming model
LOS	line of sight
MDA	Milestone Decision Authority
M-DM	mounted-dismounted maneuver
MIC	marginal implementation cost
MOMC	marginal operating and maintenance cost
MPC	marginal procurement cost
MRLCC	marginal remaining lifecycle cost
MS	maneuver sustainment
MSp	maneuver support
MUGC	marginal upgrade cost
NLOS	non–line of sight
P	protection
PE	program element
POM	program objective memorandum
PortMan	(RAND's) portfolio analysis and management method
R&D	research and development

RLCC	remaining lifecycle cost
RSTC	remaining S&T cost
RV	required value
S&T	Science and Technology
SRD	strategic responsiveness and deployability
TEL	training, leadership, and education
TEV	total expected value
TMIC	total marginal implementation cost
TRADOC	Training and Doctrine Command
TRLCC	total remaining lifecycle cost
TRSTC	total remaining S&T cost
UAV	unmanned air vehicle
USD (AT&L)	Under Secretary of Defense for Acquisition, Technology, and Logistics
WTC	Warfighting Technical Council

CHAPTER ONE

Introduction

Department of Defense (DoD) acquisition policy since 2003 has stipulated that an analysis of alternatives (AoA), including lifecycle costs, be conducted for major acquisition programs. The AoA is performed at the concept refinement stage and is due at Milestone A upon entrance into the technology development stage. Although DoD's acquisitions of systems and Science and Technology (S&T) programs have always been connected, in 2008, for the first time, DoD explicitly indicated that technologies coming out of the S&T programs can be inserted directly into every stage of the acquisition process upon the approval of the Milestone Decision Authority (MDA) (Figure 1.1).[1] This direct insertion makes it important to consider lifecycle cost at the S&T stage, because it is best to conduct a trade-off between performance and lifecycle cost before the system design concept has been finalized. Since the Army S&T community has not developed a method to accomplish such trade-offs, this study, as well as its predecessor,[2] aims to develop a methodology to incorporate lifecycle cost into S&T planning. The concept-refinement stage has now been replaced with materiel solution analysis, as shown in Figure 1.1. There are four options for S&T programs to enter into the acquisition process. Option 1 is the oldest way. The system concept gained from the S&T program is used in the design of a new or upgraded system. While TAS-1 urges lifecycle consideration during the S&T phase, under Option 1, planners at least have a second chance to consider lifecycle cost at the materiel solution analysis stage and, in the event of a major program, when an AoA is performed. It is important to note that no second chance exists for the other three options. Option 2 is to insert the S&T results into a system during technology development, when the system development is already ongoing and the time to do the initial AoA has already passed.[3] Option 3 is to transition the system or its module that resulted from the S&T program into engineer-

[1] While such direct insertions into any stage did occur prior to 2008, the 2008 DoD acquisition policy encourages them.

[2] The predecessor—Chow, Silberglitt, and Hiromoto, 2009 (hereafter referred to as TAS-1)—provides the basic model. The current study (TAS-2) adds a simulation model to deal with uncertainties in S&T project success.

[3] While the initial AoA can be updated as necessary at Milestones B and C, a system or its module developed under an S&T program, but without consideration of lifecycle cost, would be lacking an initial estimate of life-

Figure 1.1
The S&T Linkage to the Defense Acquisition Management System

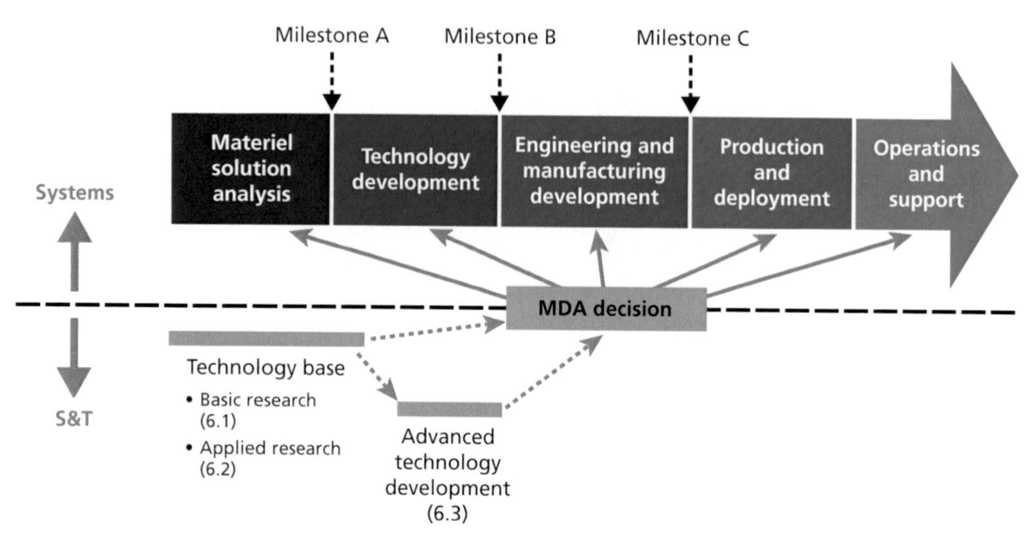

SOURCE: Simplified graph from Brown, 2008.
RAND *MG979-1.1*

ing and manufacturing development (EMD). If lifecycle consideration is already captured during the S&T stage, it would facilitate the consideration and adaptation of the S&T products into the developing systems. Option 4 is to adapt modules developed under S&T into systems already in production or even fielded. It would be too late to tailor the design of a module for affordability. In other words, it would be too late to allow for performance-cost trade-offs on the module. In sum, while the 2003 update of the DoD acquisition policy moved the lifecycle cost consideration earlier in the process, to Milestone A, the 2008 update encourages transitioning S&T programs directly into system acquisition at every stage, which necessitates consideration of lifecycle costs even earlier—at the S&T stage.

DoD Capability Portfolio Management and Cost Assessment

In September 2006, Deputy Secretary of Defense Gordon England issued a memorandum requesting experimentation with capability portfolio management (CPMT) for planning and implementing capability development. This experimentation started with four capability portfolios of like capabilities (England, 2006). These portfo-

cycle cost for an update. More important, the performance–lifecycle cost trade-off could be restricted or even too late.

lios were expected to be a part of the FYs 2008–2013 program-review process. Each capability portfolio manager (CPMR) was expected to generate a capability portfolio program-objective-memorandum assessment for the deputy's advisory working group, which would address how the current portfolio would align with "strategic interests" and "warfightings' needs." Further, the assessment would provide the "optimum mix of capabilities," which requires "minimizing gaps and overlaps" and delineating "the dependencies on other portfolios." Finally, it would recommend trades within the "portfolio to achieve the optimal mix." For the following FYs 2009–2013 program-review process, each CPMR was expected to recommend "portfolio-specific guidance" to "the components with respect to their capability portfolio." Also, "to promote transparency," the CPMRs were expected to "actively participate in the development of component POMs [program objective memorandums] related to their capability areas," and they "could request to lead the coordination of all programmatic issues" and "deliver an integrated capability portfolio–focused issue paper to the Program Review process leads." Unfortunately, the CPMRs have been unable to accomplish these tasks because they did not have the necessary processes and tools. The process, including tools, developed in our study and its predecessor, TAS-1, is a CPMT and can be used to address the issues mandated by the memorandum. Our two reports (TAS-1 and TAS-2) describe our CPMT's applications to the Army's S&T programs, while an ongoing study addresses our CPMT's applications to the Army's Engineering and Manufacturing Development programs. Further, our CPMT can also be applied directly to a portfolio of like capabilities that a CPMR is charged to address. In February 2008, Deputy Secretary of Defense England formalized the first four as standing CPMRs and added five more for further experimentation. Our process and tools are aimed to contribute to an effective CPMT for these and any future CPMRs for the Army and the other services.

On May 22, 2009, President Barack Obama signed the Weapon Systems Acquisition Reform Act into law to reform the Pentagon's process in developing and buying major defense acquisition systems. The act created a position for the Director of Independent Cost Assessment to conduct independent estimates of the cost of new major military systems. The act also states the following:

> The Secretary of Defense shall ensure that mechanisms are developed and implemented to require consideration of trade-offs among cost, schedule, and performance objectives as part of the process for developing requirements for Department of Defense acquisition programs (P.L. 111-23).

Because it is best for the trade-offs to take place at the S&T stage before the system design concept is finalized, the process and tools developed in this study and TAS-1 can be used to assess these trade-offs.

Incorporating Cost Assessment into the Army's S&T Review Process

In TAS-1, we suggested an iterative procedure for making and updating lifecycle cost estimates within the Army's annual S&T review process. The technology base consists of basic research (6.1) and applied research (6.2) (see Figure 1.1). Upon review and decision by the oversight panel and the annual S&T review process as discussed in TAS-1, selected projects will proceed to advanced technology development (6.3), with the more mature and important projects proceeding to a demonstration in an operational environment to evaluate the technology's military utility. This demonstration can take place via an Advanced Concept Technology Demonstration (ACTD), joint capability technology demonstration (JCTD), advanced technology demonstration (ATD) or by taking part in laboratory or field demonstration and warfighting experiments. Again, the oversight panel and the annual review will nominate some for an MDA decision on whether to enter into the acquisition process.[4]

It is especially important to consider lifecycle cost for the Army Technology Objectives (ATOs), which Headquarters Department of the Army designates as the highest-priority efforts within the 6.2 and 6.3 programs (U.S. Army, 2007, p. I-8). As ATOs are intended to mature technologies and transition them to program managers for system acquisition, lifecycle cost is a critical factor in system design during the ATO stage. In other words, it would be more efficient to consider affordability during the ATO stage so that performance-cost trade-offs can be made early in the design process. The alternative of modifying basic designs after the programs enter into acquisition is likely to be more difficult, take longer, cost more, and entail greater risk.

Under a previous study, we applied an expanded version of PortMan, RAND's research and development (R&D) portfolio analysis and management method (Silberglitt and Sherry, 2002; Silberglitt et al., 2004), to allow for the consideration of lifecycle cost in ATOs. This included developing a linear programming model (LPM) so that the total remaining S&T cost (TRSTC)[5] and the individual force operating capability (FOC) requirements could be conveniently added as model constraints. The objective of this LPM is to select and fund a subset of current ATOs so that the total remaining lifecycle cost (TRLCC)[6] is minimized, while all the model constraints are satisfied. This monograph describes the results of a follow-on study that allowed the

---

[4] For major defense acquisition programs under the Army, the MDA is the service acquisition executive, the Assistant Secretary of the Army (Acquisition, Logistics, and Technology).

[5] The total remaining S&T cost is the future S&T cost required to complete the S&T projects (ATOs) selected for the portfolio. It is part of the total remaining lifecycle cost.

[6] Total remaining lifecycle cost is (1) the future lifecycle cost that still has to be paid in order to complete the selected ATOs (i.e., the total remaining S&T cost) and to develop and demonstrate the new systems derived from the ATOs and (2) the difference between (a) the costs of acquiring units of these new systems over a 20-year period and operating and maintaining them over their lifetimes and (b) the costs of the legacy systems. The total does not include past lifecycle cost, which is already spent and should not enter into future decisions.

LPM to consider uncertainty. While this monograph aims to be self-contained, readers who are interested in the details of the method are recommended to also review the report of the previous study (TAS-1).

Method to Yield Affordable and Robust Systems

During this study, we developed a method, including a LPM and a simulation, to accomplish two aims: (1) provide an objective means to perform quantitative what-if analyses and (2) serve as a means to optimize a factor of the user's choice. Specifically, these tools can be used to accomplish the following:

- Compare alternative R&D strategic plans according to the degree to which they meet capability requirements, their required R&D budgets, and their total life-cycle costs.
- Perform trade-off studies of performance versus cost.
- Perform sensitivity analyses.

As an optimization method, the tools can be used to do the following:

- Prioritize R&D investments according to different objective functions or factors to optimize. For example, planners can choose to minimize total lifecycle cost or total S&T budget, or to maximize the performance of the S&T portfolio.
- Determine the percentage of the total lifecycle cost that should be allocated to S&T or EMD, as opposed to procurement, operations, and support.

Study Objective

In our previous project (as described in TAS-1), we expanded RAND's PortMan portfolio analysis and management method to include the consideration of lifecycle cost and to require the ATOs to meet individual capability requirements. In this project, we added a simulation to determine the implications of ATOs whose probability of meeting their performance and cost goals is not 100 percent. This monograph describes and demonstrates our process, which includes the LPM and the simulation, through various applications that seek to facilitate improved selection and management of the Army's ATO portfolio.

We used the ATOs in the 2007 *Army Science & Technology Master Plan* (U.S. Army, 2007), which was the latest available during this study. Accordingly, we used capability gaps of similar vintage identified by the Training and Doctrine Command (TRADOC)/Army Capabilities Integration Center (ARCIC) in data from 2006 in an unpublished report on force capability-gap analysis. However, since we lacked the detailed data on gaps and ATOs necessary for estimation of the input parameters that are accurate enough to inform decisions about actual S&T projects, one should not draw any conclusions from this monograph about the merits or drawbacks of any specific S&T projects that were chosen in this study for demonstration purposes. It should be emphasized that the purpose of this study is to develop and demonstrate a methodology, not actual applications of the methodology.

Report Structure

Chapter Two is a description of the expanded method developed during this study. Chapter Three applies the gap-space method as if it were used for decisionmaking on the FY 2007 plan during the summer of 2006, as we used only data available at that time. Chapter Four describes various applications to the gaps and ATOs of 2007, with the aim of improving ATO portfolio management. Chapter Five summarizes this study's findings and recommendations.

There are two appendixes that provide details on technical aspects of the study. Appendix A provides estimates of expected values (EVs) for individual ATOs and gaps.[7] Appendix B provides estimates of marginal implementation cost (MIC) for systems derivable from ATOs and ACTDs using a surrogate Delphi method.[8]

---

[7] As already noted, these EV estimates are for demonstration of the methodology only and are not intended to reflect actual values.

[8] Marginal implementation cost plus the remaining S&T cost equals the remaining total lifecycle cost. A more detailed description is provided in Chapter Two.

CHAPTER TWO

The Expanded Method Including Models

The RAND PortMan R&D portfolio analysis and management method has been expanded since its inception in 2002. In this chapter, we first introduce two definitions and summarize the salient features of the PortMan expansions developed during our previous study, as described in TAS-1. Then we describe the work performed under the current study, together with its associated expansions of PortMan. This description begins with a discussion of the underlying principle of gap analysis, measuring both the marginal capability contribution and cost of ATOs relative to those already associated with the existing or legacy systems. We then describe the gap-space method for estimating an ATO's contribution to capability gaps. Next is a discussion of a surrogate Delphi method[1] for estimating the cost to implement[2] the systems in order to meet the capability gaps. This exercise provides the basis for the use of a Delphi method for this purpose in the future, as the Delphi method is much quicker and much less labor-intensive to carry out than the detailed, bottom-up estimate of individual cost components used in the previous study and described in TAS-1. This chapter closes with a description of the development of a simulation for handling uncertainties regarding project success.

Definition of Terms

We begin with the following basic definitions:

---

[1] Because the purpose of this study is to demonstrate the method, not to analyze an actual case and make recommendations, we performed a simplified Delphi exercise internally, as opposed to an actual one. Our results are described in Appendix B.

[2] The implementation cost includes all remaining lifecycle cost except that of completing the S&T project: the engineering and manufacturing cost, the acquisition cost, the upgrade cost, the operating and maintenance cost, and the disposal cost (see the section "Bottom-Up Cost Estimation" in this chapter for a more detailed discussion). However, for the purpose of methodology demonstration, this study does not consider disposal costs and assumes them to be zero.

Definition 1

The contribution of the ATO i to capability gap k of FOC j is defined as $V_{i,j,k}$. $V_{i,j,k}$ is a random variable whose randomness is described by a probability distribution, $P_s(V_{i,j,k})$. Since the contribution of the ATO i can end up at a number of states where $s = 1, 2 \ldots S$, the expected value of $V_{i,j,k}$ is

$$\mathrm{E}\left[V_{i,j,k}\right] = \sum_{s=1}^{S} P_s\left(V_{i,j,k}\right) V_{i,j,k}.$$

Further, The contribution of ATO i to FOC j is the value, $V_{i,j}$, which is the sum over m gaps,

$$\mathrm{E}\left[V_{i,j}\right] = \sum_{k=1}^{m} \mathrm{E}\left[v_{i,j,k}\right].$$

Definition 2

The required value (RV) for FOC j is defined as RV_j. This is the value that the Army requires all of the funded ATOs taken together to contribute to FOC j.

With these definitions, meeting or exceeding an FOC requirement can be expressed by the following equation:

$$\sum_{i=1}^{n} x_i \mathrm{E}\left[V_{i,j}\right] \geq \mathrm{RV}_j \text{ for } j=1, 2, \ldots, 11,$$

where

x_i = 0 or 1, for an ATO that is not included or included, respectively, in the selected portfolio and

n = the number of ATOs (75 in this study).

This relationship allows users of this study's method to define a required value (RV_j) for *each* FOC and, thus, to select a project portfolio (i.e., a selected group of ATOs) that simultaneously meets all of these required values. Users can also change the requirements to see how the change affects the portfolio selection.

Since some studies may use a Delphi method to estimate the $\mathrm{E}[\mathrm{V}_{i,j}]$ directly, it can also be called the EV score for the *i*th ATO on the *j*th FOC, or simply the $\mathrm{EV}_{i,j}$ score:

$$\mathrm{EV}_{i,j} = \mathrm{E}[\mathrm{V}_{i,j}].$$

Another useful parameter is the total EV score from all ATOs on the *j*th FOC, or simply the EV_j score:

$$EV_j = E\left[V_j\right] = \sum_{i=1}^{n} E\left[V_{i,j}\right].$$

To estimate EV scores for ATOs, this study uses ATO data from the *Army Science & Technology Master Plan* (U.S. Army, 2007) and gap data from 2006 in an unpublished report from TRADOC/ARCIC on force-capability gap analysis.

Key Features of the Previous Study and This Study

TAS-1 added three key features to the PortMan method. First, the team developed a gap-space analysis to estimate the contributions of ATOs to individual FOC gaps. Second, it developed a detailed bottom-up method to estimate the cost components of the total remaining lifecycle cost. Third, it developed an LPM to arrive at the lowest total remaining lifecycle cost. However, this was a "certainty model," since any ATO was assumed to be successfully completed on time and on budget and to ultimately attain its performance and cost goals if the ATO were funded until completion. The following sections describe how the first two features are applied in this monograph. The last subsection on simulation describes how the added simulation model can be used with the LPM to account for the fact that some ATOs will inevitably fail to be successfully completed and to meet their original performance and acquisition and/or fielding cost goals.

Gap-Space Method for Estimating Value

We used the gap-space-coverage method developed under the previous study and described in TAS-1 for estimating value. However, in the previous study, we used the FOC sub-requirements[3] as a surrogate for the capability gaps identified by TRADOC for each FOC. For our current study, we adopted the actual capability gaps as described in 2006 data from TRADOC/ARCIC on force-capability-gap analysis.

Gap-Space Coverage

We begin by briefly reviewing the gap-space-coverage method of the previous study. By *gap space* we mean all of the TRADOC/ARCIC 2006 current force capability gaps, or FOC gaps.[4] We then define the *gap-space coverage* of any ATO in terms of the gaps

[3] For example, protect personnel, assets, and information are listed as the three sub-requirements of the FOC on protection.

[4] After we combined several pairs of almost identical gaps and a few projects with their follow-on projects, the resulting number of ARCIC capability gaps was 113. Then, we grouped these into the 11 FOC categories. These

(or portion of the gap space) addressed by that ATO. A gap is measured in units of its desired capability, e.g., speed or bandwidth of communication. An ATO may address a gap completely or partially, e.g., improved bandwidth at the same bit rate. To estimate the gap-space coverage of an ATO, we need a way to match the capabilities needed to address each gap with the capabilities that will be supplied if the ATO is successful. We begin by recognizing that FOCs apply to warfighters in three different situations: (1) off the battlefield, (2) on the way to the battlefield, and (3) on the battlefield, as indicated in Figure 2.1. Then we ask where the capability gaps apply to each of these situations and where capabilities that will be supplied by successful ATOs apply to each of these situations. Next we divide the FOC gap space into a mutually exclusive set of categories tailored to each FOC, as indicated in Table 2.1. Within each of these categories, a capability gap will be measured in units of its required capabilities, and these required capabilities can be compared with the capabilities that will be supplied by successful ATOs.

According to these definitions, each FOC capability gap will occupy a portion of this gap space that is determined by the situations and categories to which it applies. Similarly, each ATO will address a portion of the gap space that is determined by the FOCs, gaps, situations, and categories that it addresses. Then, for each gap that an ATO addresses, we define its coverage of that gap by the fraction of the situations and categories to which that gap applies that are also addressed by the ATO. Finally, we estimate the gap-space coverage of the ATO for each FOC by adding together all the contributions to each gap and dividing by the number of TRADOC/ARCIC gaps we grouped under this FOC, as indicated by the equation below:

$$\mathrm{CV}_{ij} = \frac{\sum_{k=1}^{n} \frac{S_{ik} \times C_{ik} \times G_{ik}}{S_{jk} \times C_{jk}}}{n},$$

where CV_{ij} is the gap-space coverage of FOC j by ATO i, which can take on any value between zero (ATO i fails to address any FOC j gaps) and unity (ATO i addresses the situations and categories of all FOC j gaps). There are n TRADOC/ARCIC capability gaps grouped under FOC j. G_{ik} is 1 for each of these n gaps that ATO i addresses and zero otherwise. S_{jk} and C_{jk} are the number of situations from Figure 2.1 and categories from Table 2.1, respectively, to which the kth gap applies, and S_{ik} and C_{ik}, when they are multiplied by G_{ik}, are the number of these situations and categories that ATO i addresses.

For example, suppose FOC_j was FOC 1, battle command, and gap k was an assumed inability to communicate with soldiers who were either on the battlefield or

FOCs are described in U.S. Army, 2005. After the grouping, we adopted the convention of calling TRADOC/ARCIC gaps *FOC gaps*. Furthermore, we call those under FOC 1 *FOC 1 gaps*, and so forth for the other FOCs.

Figure 2.1
Situations in Which FOCs Apply to Warfighters

RAND *MG979-2.1*

on their way to the battlefield. Further, suppose ATO_i was a communications system that was effective only at short range on the battlefield. For this fictional example, S_{1k} would be 2 (situations 2 and 3 of Figure 2.1), C_{1k} would be 1 (communicate, one of the four categories under FOC 1, battle command in Table 2.1); S_{ik} would be 1 (situation 3 of Figure 2.1), and C_{ik} would be 1 (communicate, one of the four categories under FOC 1, battle command in Table 2.1). Thus, for this fictional example, the contribution of ATO_i for gap k to CV_{i1} would be 1 × ½ × 1 = 0.5. We note that this assumed ATO_i might also address other gaps under FOC 1, so those contributions must be added, and the total then divided by n to obtain CV_{i1} for this assumed ATO_i. Furthermore, other G_{ik} for this assumed ATO_i may be nonzero, as it might address gaps in other FOCs, e.g., in the equipment and supplies category for FOCs 3, 4, 6, and 9 in Table 2.1, and accordingly might also make contributions to CV_{ij} for other FOCs.

The ATOs that we used in this study were those described in the Army S&T master plan (U.S. Army, 2007). We assume that this set of ATOs funded in fiscal year (FY) 2007 constitutes the full set to address the current force residual gaps[5] identified by TRADOC/ARCIC in May 2006. On the other hand, we note that the ATOs are only a portion of the S&T program, i.e., those identified as "the most important S&T programs" aimed at filling FOC gaps. Our study does not address the basic research (6.1) program, and the non-ATO S&T projects in applied research (6.2) and advanced technology development (6.3) that, together, may provide additional capabilities to address longer-term FOC gaps and other Army needs.[6]

[5] Residual gaps are longer-term gaps, because they are the remaining gaps that the completed ATOs and those in Engineering and Manufacturing Development still cannot fill. These remaining gaps will have to wait for ongoing and future S&T projects to be met.

[6] This study is meant to demonstrate a methodology. In future applications for actual decisionmaking, planners can include all projects that address the gaps. Our methodology can be generalized to accommodate programs that aim to develop specific systems (as in the focus of this study) and also 6.1, 6.2, and other programs that contribute broadly to future capabilities and generally to multiple systems.

Table 2.1
Categories of FOC Gap Space

FOC	Basis for Categories	Categories
1. Battle command	Must provide	Command, control, communications, and computers
2. Battlespace awareness	Must provide common operating picture	LOS fire, B/NLOS fire, force location, and hazard location
3. Mounted-dismounted maneuver	Needed for maneuver	Forces, mobility, weapons, and equipment and supplies
4. Air maneuver	Needed for maneuver	Forces, mobility, weapons, and equipment and supplies
5. LOS/BLOS/NLOS lethality	TRADOC defined	LOS, BLOS, NLOS
6. Maneuver support	Needed for maneuver	Forces, mobility, weapons, and equipment and supplies
7. Protection	Must protect	Personnel against disease[a] and injury[b], assets[b], and information[c]
8. Strategic responsiveness and deployability	Must provide	Readiness, transportation, and delivery
9. Maneuver sustainment	Needed for maneuver	Forces, mobility, weapons, and equipment and supplies
10. Training, leadership, and education	Must train or educate concerning	Doctrine, equipment, people, and environment
11. Human engineering	Must engineer[d]	Systems, people or tasks, and human-system interfaces

[a] For disease, the categories are malaria; HIV/AIDS; tuberculosis; emerging infectious diseases, e.g., dengue fever; and influenza.

[b] For injury and asset protection, the categories are accident, line of sight (LOS) fire, B/NLOS (beyond and non–line of sight) fire, and hazards (e.g., mines, booby-traps, and improvised explosive devices [IEDs]).

[c] For information protection, the categories are theft, destruction, modification, and access restriction.

[d] If the engineering is to reduce soldier load (e.g., see U.S. Army 2005, paragraph 4-73), then the categories are (a) food, (b) clothing and shelter, (c) weapons, and (d) equipment and supplies.

To convert our FOC gap-space-coverage estimates for ATOs into estimates of their expected values, we require data that will allow a measure of the degree to which the anticipated payoffs of the ATO are actually achieved based on the performance levels of the systems developed from the ATO. In other words, what fraction of the gap—measured in units of the required capabilities (speed and bandwidth of communication in the example described at the beginning of this subsection)—will the systems developed from the ATO actually achieve. In the RAND PortMan method, a scaling factor is provided with a value between zero and 1 that is called the *technical potential* (Silberglitt and Sherry, 2002; Silberglitt et al., 2004). Within the portfolio analysis approach that was used by the Assistant Secretary of the Army (Acquisition, Logistics and Technology), an expert panel estimated the *technical feasibility* of each ATO. This technical "feasibility" could be used as a surrogate estimate of the PortMan technical "potential." However, since we do not have access to these data or any other validated surrogates, for this demonstration, we assume an equal value of one-half for the *technical potential* of each ATO, so that our expected value estimate of ATO i for FOC j is given by:

$$EV_{ij} = \frac{1}{2} \times CV_{ij}.$$

The data that we used to estimate the ATO CVs and EVs were derived from the Army S&T master plan (U.S. Army, 2007) and an unpublished TRADOC/ARCIC report on force-capability-gap analysis using 2006 data. Chapter Three provides a description and analysis of the gap-space coverage matrix for all ATOs and FOCs, as well as numerical estimates of coverage by ATO and gap and the distribution of ATO EVs by FOC. Expected value estimates by ATO and gap are presented in Appendix A.

Bottom-Up Cost Estimation

Recall that our study addresses how ATOs can meet capability gaps, which represent needed improvements in mission performance of Army systems. Therefore, costs should also be measured in a marginal sense. We start with a legacy baseline, which consists of systems already in service and the acquisition of additional (unimproved) legacy systems, but without any improved or new systems. The capability provided by these aggregated legacy systems is called the *baseline capability.* In this manner, the capability gaps are the Army's remaining required capabilities that existing and additional legacy systems cannot meet. Similarly, the costs of meeting the capability gaps are those that are above the corresponding costs for the (legacy) baseline capability. For example, if the total cost of acquiring and servicing the legacy systems is $100 billion and if replacing some legacy systems and adding some new systems would cause a net increase in total cost of $30 billion, the marginal cost for filling the capability gaps would be $30 billion. This formulation allows negative marginal cost if the new system can replace the legacy system at a lower cost. This feature of our methodology is important, because the contributions from ATOs that aim to save implementation cost are often not properly captured using current methods. For example, the dollar savings, if captured at all, might be commingled with capability performance improvements. These two contributions should not be mixed, as they are measured in different units.[7]

[7] It should be noted that all ATO contributions are independent and additive. For example, assume that there is an ATO project that can improve the capability of a legacy system by adding a component to it. There is another

We broke the cost into its key components, as follows:

- Remaining S&T cost (RSTC) is the future cost to complete the ATO program.
- Engineering and manufacturing development cost (EMDC), which was previously called system development and demonstration cost, is the cost to develop and demonstrate the new system derived from the ATO in question.[8]
- Marginal procurement cost (MPC) is the cost of acquiring new systems, minus the cost of buying legacy systems instead, to serve the planning period. It is the number of units procured times the marginal unit cost. The latter is the unit cost of a new system, minus that of the legacy system to be replaced.
- Marginal upgrade cost (MUGC) is the cost to modify the new system to maintain required performance, minus that required for the legacy system.
- Marginal operating and maintenance cost (MOMC) is the cost difference between servicing the new systems over the planning period versus servicing the legacy systems. For some systems, a major contributor to MOMC is the marginal manpower cost, which is the manpower cost for operating the new system during its operating life over that of the legacy system. The manpower operating cost is equal to the annual manpower cost times the operating life. The annual manpower cost is personnel time spent in operating the system during the year times the salary rate.

All five terms together constitute the marginal remaining lifecycle cost (MRLCC), which is the cost difference between the new system's remaining lifecycle cost (RLCC) and that of the legacy system. Often, MIC is used to represent the last four terms.

Thus,

$$\text{MRLCC} = \text{RSTC} + \text{MIC},$$

where

$$\text{MIC} = \text{EMDC} + \text{MPC} + \text{MUGC} + \text{MOMC}.$$

ATO project that can lower the cost of the legacy system by using cheaper components but does not lower the capability. If both ATOs are successful and are applied to the legacy system and result in 100 percent of both benefits (improved capability and lower cost), they are independent and additive and can be included in the model. On the other hand, if, when both projects are applied, the benefits are less than 100 percent for either one, we can add a constraint in the LPM to account for the interdependence between the two ATOs. See TAS-1, pp. 26–27.

[8] The S&T project is assumed to be transitioned into the Defense Acquisition Management System as shown in Figure 1.1 at the EMD phase. If the project is transitioned into a different phase, the cost is adjusted accordingly.

There is also a discount rate,[9] which is the interest rate to discount future costs so as to be comparable with the current costs.

In TAS-1, we estimated these cost components for each ATO by comparing them with the historic costs of similar legacy systems or by using scenarios to determine the number of units required to meet future capability gaps. In TAS-1, we also converted the marginal implementation costs into seven grading levels.[10] Although these cost levels are a very rough approximation of the detailed bottom-up cost estimates, we found in TAS-1's Appendix F that model selection of ATOs for the optimal portfolio is similar whether the detailed bottom-up costs or the rough grading cost levels are used. For simplicity in the present study, we used a surrogate Delphi method with these grading levels, as described in Appendix B, to estimate the marginal implementation costs. On the other hand, if Army offices have detailed bottom-up cost estimates from sources such as the AoA, they can use such cost estimates instead.

Linear Programming Model

The LPM selects a package of ATO projects such that the cost to complete these selected ATO projects and to develop, field, and operate their resulting systems to meet all FOC requirements is minimized. This can be expressed as follows:

Minimize

$$\sum_{i=1}^{n} x_i \mathrm{MRLCC}_i$$

subject to a constraint

$$\sum_{i=1}^{n} x_i \mathrm{RSTC}_i \leq \mathrm{BSTC}$$

and a set of 11 constraints

$$\sum_{i=1}^{n} x_i \mathrm{E}\left[V_{i,j}\right] \geq \mathrm{RV}_j \text{ for } j \text{ from 1 to 11 (FOCs),}$$

where

x_i = 0 or 1 for nonselected and selected ATOs, respectively, where i is the ATO program number running from 1 to n (75 in this study),

MRLCC_i = the marginal remaining lifecycle cost, which includes RSTC_i, for the systems resulting from ATO i,

[9] We assume zero-percent discounting in TAS-1 and here, as both studies are demonstrations of the method. On the other hand, a 3-percent real discount rate (after inflation) is typically used by the Department of Defense.

[10] See Table 2.2 in this monograph.

$RSTC_i$	=	the remaining S&T cost for ATO i,
$BSTC$	=	the budgeted total remaining S&T cost for all the selected ATO projects,
RV_j	=	the required value for FOC j.

Expansions Under the Current Study

This study added two modifications and one expansion to the version of PortMan described in TAS-1. First, as described above, we refined the gap-space analysis by applying it directly to the capability gaps identified by TRADOC/ARCIC. This analysis is detailed in Chapter Three and in Appendix A. Second, since we have already demonstrated the detailed bottom-up method in TAS-1, this current study focuses instead on the demonstration of a Delphi method for cost estimation, which is discussed in the next subsection and in Appendix B. Third, the study team developed a simulation to analyze the implications of uncertainties in the likelihood of success of ATOs. This analysis is discussed in the last section of this chapter.

Surrogate Delphi Method

Since this study is a demonstration of a method and applications, we did not seek the expert participants necessary to carry out an actual Delphi exercise, but rather used two study team members to simulate the Delphi exercise for the estimation of ATO implementation costs. We called our estimation, for demonstration purposes only, a surrogate Delphi method. For these cost estimates, we used the cost ranges that would be used in the actual Delphi, but study team analysts instead produced the estimates. These cost estimates were then used in a simulation that addressed the effects of uncertainties on the selection of the ATO portfolio to meet all of the individual FOC requirements within the constraints of the total remaining S&T budget and the total remaining lifecycle cost.

As described in the section above, "Bottom-Up Cost Estimation," the remaining lifecycle cost can be divided into two components: the remaining S&T cost and the marginal implementation cost. This cost division matches well with the needs of ATO portfolio selection and management. For example, the selection of a subset of existing ATOs for continued funding to completion with the objective of maximizing the probability that all the capability gaps will be met, subject to a constraint on total remaining S&T budget, is based on two decisions. The first decision is to select a subset of existing ATOs for continuing funding to completion, whose cost is the remaining S&T cost. As this study assumes that each ATO will have a 90-percent chance of meeting the performance and cost goals originally set forth, some of these selected

ATOs will fail.[11] The second or subsequent decision is to select a subset of the successful ATOs for implementation such that the systems derivable from these ATOs will be procured, fielded, operated, and maintained to meet all the capability gaps.[12] The cost of this second decision is the marginal implementation cost. The next subsection on simulation will further describe the decision process. In this subsection, we focus on the estimation method for the remaining S&T cost and the marginal implementation cost.

While the annual Army S&T master plan describes each ATO and ACTD,[13] it does not provide any cost data. Moreover, such cost data are neither publicly available nor easily obtainable. The cost data available to this study was that from Army program elements (PEs). However, a PE typically contains sub-elements pertaining to multiple ATOs. We used our best judgment to segregate the cost of each PE into components for each of these ATOs. The remaining S&T costs for each ATO were then determined as the sum of these cost components from all of the PEs with which the ATO is associated (see Figure 2.2).[14]

To estimate the marginal implementation cost of the systems derived from a successful ATO, two study team members were chosen as cost estimators, taking the roles that would be filled by participant experts in an actual Delphi exercise. Following Appendix F of TAS-1, the cost estimators were presented with the following definitions of each grade:

- Grade –2 is for an ATO whose description indicates that its objective is to significantly reduce the implementation cost by more than $300 million of marginal cost.
- Grade –1 is for an ATO whose description indicates that its objective is to slightly reduce the implementation cost of legacy systems by up to $300 million.

---

[11] The simulation allows individual ATOs to have different probabilities of success. For a demonstration in this study, we assume that the historic overall success rate of past ATOs is 90 percent. Further, we assume that planners cannot tell in advance which ongoing ATO is riskier than others. Then, it is reasonable to assign the same historic overall success rate to every ongoing ATO in advance of their completion.

[12] This study uses the simplifying assumption that all ATOs are completed at the same time. This assumption may be relaxed either via an approximation such as choosing the decision point to coincide with the time that the last ATO is completed or an exact approach that uses a sequence of decisions. For the sequence approach, the first decision point occurs when the first ATO is complete. Planners would then reassess the probabilities of success of all the other ATOs and select a new subset of ATOs for continued funding. This process is repeated upon the completion of the first ATO of the new subset and repeated again and again until the last ATO is completed.

[13] Our set of S&T projects includes ATOs and ACTDs. For convenience, the term ATO is hereafter used to include ACTDs as well.

[14] As previously stated, the objective of this study is to develop a methodology for the Army and for other services' use. Should the Army decide to adopt this study's method, the problem of misallocating some costs to the wrong ATOs would not occur because the Army knows the remaining S&T cost for each ATO directly, and there is no need to derive it from the aggregated PEs.

Figure 2.2
Remaining S&T Costs for Individual ATOs

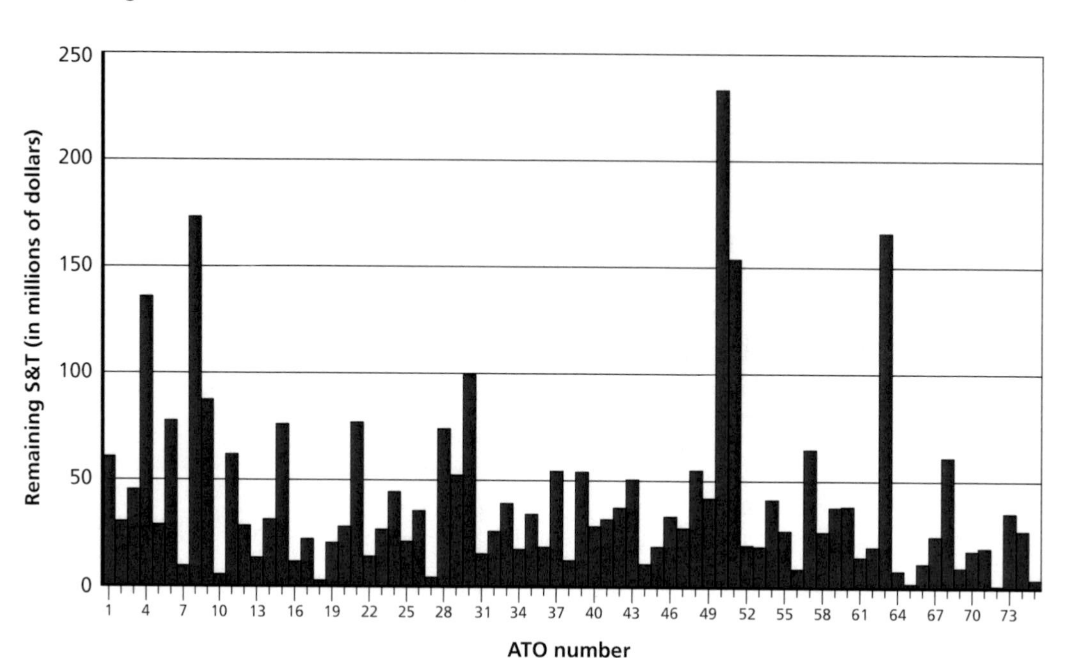

RAND *MG979-2.2*

- Grade 0 is the implementation of enough new systems to serve over the planning period that cost about the same or up to $150 million more than the legacy systems being replaced.
- Grade 1 is for an ATO with system implementation cost only slightly higher than that of the legacy system to be replaced. Alternatively, each system's cost can be considerably higher if the number of units to be implemented is small, so that the product of the two (unit cost times number of units) is between, say, $150 million and $300 million of marginal implementation cost.
- Grade 2 is for an ATO for which the unit cost of the new system is considerably higher than that of the legacy system or the number of units deployed is more numerous, so that their product is between $300 million and $2 billion of marginal implementation cost. This large range makes it rather easy to determine whether the marginal implementation cost of a particular ATO's systems belong to this grade.
- Grade 3 is for marginal implementation cost between $2 billion and $4 billion. Again, it should be feasible to decide whether a particular ATO's systems belong here.
- Grade 4 is for very large marginal implementation costs, above $4 billion.

These grade categories are summarized in Table 2.2.

Table 2.2
Numerical Grade Descriptions for Marginal Implementation Costs

Grade	Marginal Implementation Cost Range	Value Used in Conversion (millions of dollars)
–2	cost < –$300 million	–450
–1	–$300 million ≤ cost < $0 million	–150
0	$0 million ≤ cost ≤ $150 million	75
1	$150 million < cost ≤ $300 million	225
2	$300 million < cost ≤ $2 billion	1,150
3	$2 billion < cost ≤ $4 billion	3,000
4	cost > $4 billion	5,000

The evaluators made three rounds of cost estimates. The first round was an independent estimate made by each evaluator. After the evaluator reviewed the other's first-round estimate, he made his second-round estimate. In round three, the two evaluators discussed the reasons behind every disagreement in their estimates. While they were free to assign any final grade, the two evaluators for this study came to a "consensus" grade for each ATO. We emphasize that this surrogate Delphi is only a demonstration. An actual Delphi exercise may have five to ten evaluators and four rounds, with consensus developed through anonymous submissions. Alternatively, there would be an uncertainty range based on the distribution of the final grades assigned by the expert participants. Both the average grades and the uncertainty ranges can be used in the post–Delphi analysis. In this study, we use only the average grades, without the uncertainty ranges.

Appendix B provides the results of the surrogate Delphi, which estimated the marginal implementation costs in grades for the 75 ATOs used in this study.

The evaluators were asked to place only the marginal implementation cost for an ATO system into one of the seven cost bins. To use the implementation costs in our models, we need to have a dollar amount, instead of a range, represent the cost in each bin. These dollar amounts are shown in the last column of Table 2.2.[15] The marginal implementation costs for systems derived from the 75 ATOs are shown in Figure 2.3.

Interestingly, the rough nature of these grades may be useful to reflect the large uncertainties in cost estimates during the S&T phase. The wide range in each grade indicates that the actual cost can shift considerably from the cost projection made at the S&T stage as the ATO progresses and the system develops. Sensitivity analysis of

---

[15] The cost is the average cost in the range, except the ranges at both ends. For these exterior ranges, we assumed that the width of the range is the same as that of its immediate interior neighbor. For example, we assume the range for Grade 4 to be from $4 billion to $6 billion, yielding an average cost of $5 billion as shown in the last column.

Figure 2.3
Marginal Implementation Costs for Individual ATOs

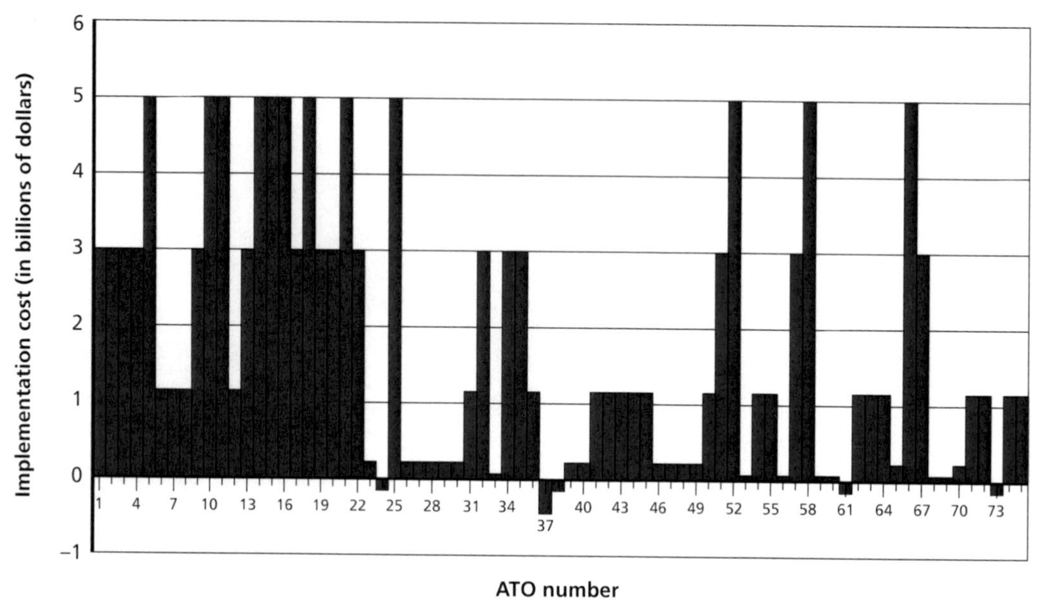

RAND *MG979-2.3*

these shifts supports the selection of ATOs that together can most likely meet future requirements for a range of uncertainties, including uncertainty in marginal implementation cost.

Simulation

The ultimate goal of our simulation is to analyze the implications of uncertainties in any number of input parameters. This study demonstrates the simulation by analyzing the uncertainties in whether the individual ATOs are successfully completed to meet their performance and cost goals. There are 75 ATOs, each with an assumed probability of success of 0.9.[16]

To understand the simulation and its objective, we need to introduce the concept of *feasible percentage*. There are at least two critical decision points in the management of the ongoing ATO portfolio. The first is the decision of which subset of the 75 ongoing ATOs should receive continued funding, especially when planners face an S&T budget cut. The second decision point occurs when the subset of ATO programs selected in the first decision is completed. With 90-percent success rates, on average, 9 out of every 10 selected ATOs are expected to be successful in meeting their goals in system performance and implementation cost. However, those that fail are random and

[16] The model can accommodate different success rates for different ATOs.

unpredictable, as they depend on the random 90-percent and/or 10-percent draw on each selected ATO. For example, in a single draw on each selected ATO, the number of successfully completed ATOs can be eight, instead of nine, out of every ten ATOs. Which ATOs will be completed successfully is also a random result. The second decision is to select systems from those successful ATOs for implementation to meet all FOC gaps. To make an optimal selection at the first decision point, planners must look forward to see the distribution of consequences of the first selection on the second decision. An optimal selection is the subset of the 75 existing ATOs that gives the highest probability of meeting all FOCs at the second decision point, within given budgetary constraints in total remaining S&T costs and total remaining lifecycle costs. For convenience of discussion, we use the term feasible percentage. The feasible percentage of a given subset of ATOs chosen during the first decision is the expected percentage of all random combinations of ATO successes and failures that fulfill all FOC gap requirements at a given budget. *In other words, for a subset of ATOs selected to be funded to their completion, the feasible percentage can be interpreted as the chance for this subset to meet all requirements within their budgets. Thus, an optimal selection is to identify the subset that has the largest feasible percentage.*

We illustrate the determination of feasible percentage with an example. First, we arbitrarily choose a set of all odd-numbered ATOs (i.e., 1, 3, 5, . . . 75). Then we randomly make a 90-percent or 10-percent success or failure draw on each of these 38 ATOs to result in a run. Based on random draws, this run will result in, say, 34 ATOs being successful and four ATOs (3, 11, 17, and 73) failing. Then, the model determines whether these 34 ATOs can meet all model constraints with a given total marginal implementation budget. If they do, they constitute a feasible run. We repeat this process for a total of 10,000 runs for the same set of all odd-numbered ATOs. If the number of feasible runs is 8,260, the feasible percentage for this set of all odd-numbered ATOs would be 82.6 percent. We then tried other sets, and the one that gives the largest feasible percentage is the optimal set, which is the set of ongoing ATOs that planners should continue funding, assuming that the budget cannot support all ATOs and the funding of some ATOs must stop. In other words, the first question for the simulation to address is *which existing ATOs should be selected for continued funding* until their completion so that the probability of fulfilling gaps in all 11 individual FOCs, or the feasible percentage, is maximized, subject to constraints on the total remaining S&T budget and the total remaining lifecycle cost.

If there were no budgetary constraints, the answer would be obvious—complete all ATOs. If the only constraint is the total remaining S&T budget, and all the probabilities of success are 100 percent, the problem has been resolved in TAS-1 using an LPM. If all the probabilities of success are not 100 percent, the problem becomes more complicated but is resolvable by the method described in Chapters Three and Four here.

CHAPTER THREE

Applications of the Gap-Space Method

This chapter explains how we applied the gap-space method to describe and analyze the degree of coverage provided for the 2006 TRADOC/ARCIC–defined current force residual FOC gaps by the Army's 2007 ATOs. It also provides numerical examples of gap-space coverage and ATO expected value estimates. We provide these gap-space method analyses and results here in a separate chapter from the applications of the full method because they may be useful in their own right in providing insights for the analysis and management of S&T projects. In performing these analyses, we imagined that we had been placed back in time to the summer of 2006. At that time, the near-term capability gaps had just been determined. The question that the method addresses is: How well would the ATOs existing at that time have met these gaps? The method identifies coverage of all individual gaps, including both extremes, namely, gaps not met by any ATOs and gaps met by many ATOs. Our analysis suggests that efficiency can be improved by terminating some ATOs pertaining to the latter group and using the saved money to fund new ATOs for the former group.

Gap-Space Coverage Matrix

The gap-space coverage matrix has a row for each ATO and a column for each FOC gap. Below the ATOs (including ACTDs as noted in Chapter Two) are listed by number and technology type as defined in the Army S&T plan and are consecutively numbered for the purposes of our analysis. The FOC gaps are indicated only by number.[1]

Force Protection Technologies

1. Network electronic warfare ATO
2. Mine and IED detection ATO
3. Rotorcraft survivability ATO
4. Kinetic Energy Active Protection System ATO

[1] An Excel spreadsheet that identifies the gaps is available with the approval of the sponsor of this study. If interested, please use the RAND contact information in the Preface of this monograph.

5. Passive Infrared Cueing System ATO
6. Extended-area protection and survivability ATO
7. Dissemination of advanced obscurants ATO
8. Pulse power for the Future Combat Systems (FCS) ATO
9. Vehicle armor technology ATO
10. Solid-state laser technology ATO
11. Countermine and IED neutralization ATO
12. Vision protection ATO
13. Modular protective systems for Future Force assets ATO
14. Wide area airborne minefield detection ATO

Intelligence, Surveillance, and Reconnaissance (ISR) Technologies

15. Third-generation infrared (IR) technologies ATO
16. Distributed Aperture System ATO
17. Suite of sense-through-the-wall systems ATO
18. Multimission radar ATO
19. All-Terrain Radar for Tactical Exploitation of Moving Target Indicator and Imaging Surveillance System ATO
20. Distributed imaging radar technology for continuous battlefield imagery ATO
21. Class-II unmanned aerial vehicle (UAV) electro-optical (EO) payloads ATO
22. Objective pilotage for utility and lift ATO
23. Soft-target exploitation and fusion ATO
24. Low-cost, high-resolution IR focal plane arrays ATO
25. Human infrastructure detection and exploitation ATO

Command, Control, Communications, and Computers (C4) Technologies

26. Tactical wireless network assurance ATO
27. Networked enabled command and control ATO
28. Tactical mobile networks ATO
29. Tactical network and communications antennas ATO
30. Battlespace terrain reasoning and awareness—battle command ATO

Lethality (L) Technologies

31. Non–line-of-sight and line-of-sight (NLOS-LS) launch system technology ATO
32. Mounted Combat System and Abrams Ammunition System technologies ATO
33. Common smart submunition ATO
34. Nonlethal payloads for personnel suppression ATO
35. Electromagnetic gun technology maturation and demonstration ATO
36. Fuze and power for advanced munitions ATO
37. Microelectromechanical systems inertial measurement unit ATO
38. Smaller, lighter, cheaper munitions components ATO

39. Hardened combined effects Penetrator warheads ATO
40. Insensitive munitions technology ATO
41. Novel energetic materials for the Future Force ATO
42. Missile propulsion technology ATO
43. Missile seeker technology ATO

Medical Technologies

44. Automated Critical Care Life Support System ATO
45. Fluid resuscitation technology to reduce injury and loss of life on the battlefield ATO
46. Vaccines and drugs to prevent and treat malaria ATO
47. Vaccines to prevent diarrhea ATO
48. Vaccine for the prevention of military HIV infection ATO
49. Biomedical enablers of operational health and performance ATO

Unmanned Systems Technologies

50. Robotics collaboration ATO and near autonomous unmanned systems ATO
51. UAV system technologies ATO
52. Army/DARPA enabling technologies for the FCS ATO
53. Manned/unmanned rotorcraft enhanced survivability ATO

Soldier Systems Technologies

54. Future Force Warrior ATD
55. Soldier mobility vision systems ATO
56. Nutritionally optimized first strike ration ATO
57. Soldier protection technologies ATO
58. Mounted/dismounted soldier power ATO
59. Infantry warrior simulation ATO
60. Leader adaptability ATO
61. Strategies to enhance retention ATO

Logistics Technologies

62. Precision airdrop–medium ATO
63. Hybrid electric for the FCS ATO
64. Advanced lightweight track ATO
65. Joint rapid airfield construction ATO
66. JP-8 reformer for alternate fuel sources ATO
67. Prognostics and diagnostics for operational readiness and condition-based maintenance ATO

Advanced Simulation Technologies

68. Learning with adaptive simulation and training ATO
69. Scaleable embedded training and mission rehearsal ATO
70. Severe trauma simulation ATO

Advanced Concept/Joint Capability Technology Demonstrations

71. Joint enabled theater access–sea ports of debarkation ACTD
72. Tactical wheeled vehicle fleet modernization and future tactical truck systems ACTD
73. Adaptive joint C4ISR node ACTD
74. Theater effects–based operations ACTD
75. Joint Modular Intermodal Distribution System JCTD.

Using the definitions of the FOC gaps provided by TRADOC/ARCIC and the definitions and data provided for the ATOs in the 2007 Army S&T master plan, we developed the full gap-space coverage matrix presented as Figures 3.1 through 3.8.[2] In those figures, red indicates that the ATO does not address the gap, and green indicates that it does. The totals in the column on the right-hand side of each FOC show the total number of gaps in that particular FOC addressed by that ATO, while the totals at the bottom of each column show the total number of ATOs that address that gap.[3] Figures 3.1 through 3.8 allow quick identification of the distribution of ATO coverage by both FOC and gap, showing which gaps and FOCs have the most and least coverage.

As indicated by Figures 3.1 through 3.8, most ATOs cover multiple gaps. Most ATOs also contribute to multiple FOCs, as shown in Figure 3.9.

To study the nature of this coverage by FOC situations and categories, which we will need to do to develop estimates of ATO coverage scores and expected values, we define the template shown in Figure 3.10. This template allows disaggregation of gaps for each FOC according to situations and categories specific to that FOC to which the gap applies.

Figures 3.11 and 3.12 summarize the distribution of gaps (the "demand" or requirement) and ATOs (the "supply" or source of technologies to fill the requirement), respectively. The numbers in the upper left-hand corners of the individual elements of Figure 3.11 show how many gaps apply to that FOC, situation, and category, while the numbers in the upper right-hand corners of the individual elements of Figure 3.12 show how many ATOs address that FOC, situation, and category. The numbers in the upper corners of the FOC boxes in the figures show the total number of gaps or ATOs

[2] In developing this matrix, we assigned each capability gap defined by TRADOC/ARCIC to the single FOC to which, in the judgment of the study team, it most closely applies. We based these judgments on the definitions of the gaps and the FOCs, including the description of FOC sub-requirements in U.S. Army, 2005.

[3] Because of the size of the gap space matrix, we show the rows for ATOs 1–38 in the first four figures, and the rows for ATOs 39–75 in the next four, so that column totals appear only in Figures 3.5–3.8.

Figure 3.1
ATOs 1–38 Gap-Space Coverage Matrix for FOC 1 Battle Command and FOC 2 Battlespace Awareness

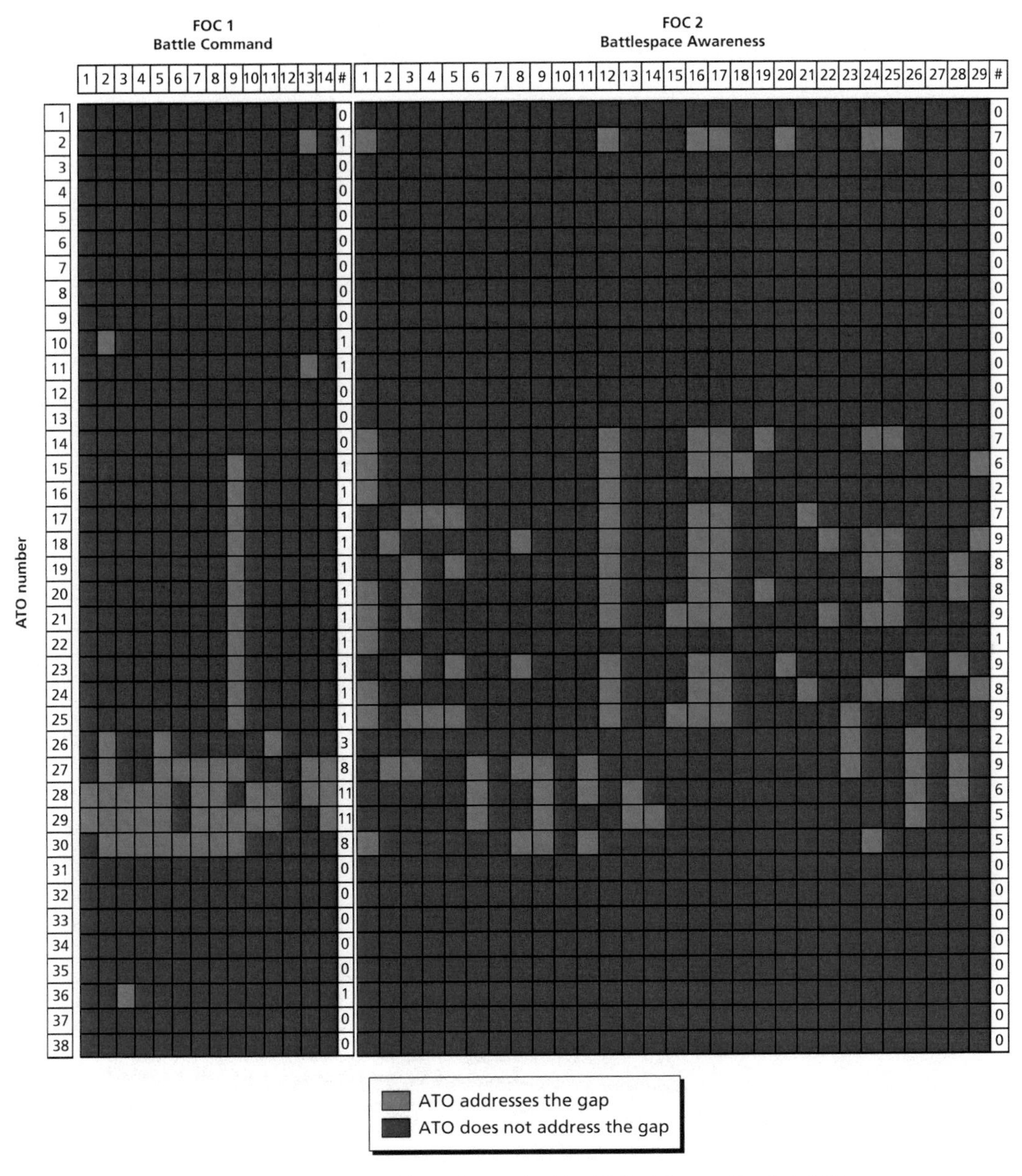

NOTE: The # column shows the total number of gaps in that FOC addressed by that ATO.
RAND *MG979-3.1*

for that FOC. The elements shaded in gray in the figures indicate FOCs, situations, and categories for which there is no gap. As shown in Figure 3.12, there are nonetheless ATOs that address these areas.

Figure 3.2
ATOs 1–38 Gap-Space Coverage Matrix for FOC 3 Mounted-Dismounted Maneuver, FOC 4 Air Maneuver, FOC 5 Lethality, and FOC 6 Maneuver Support

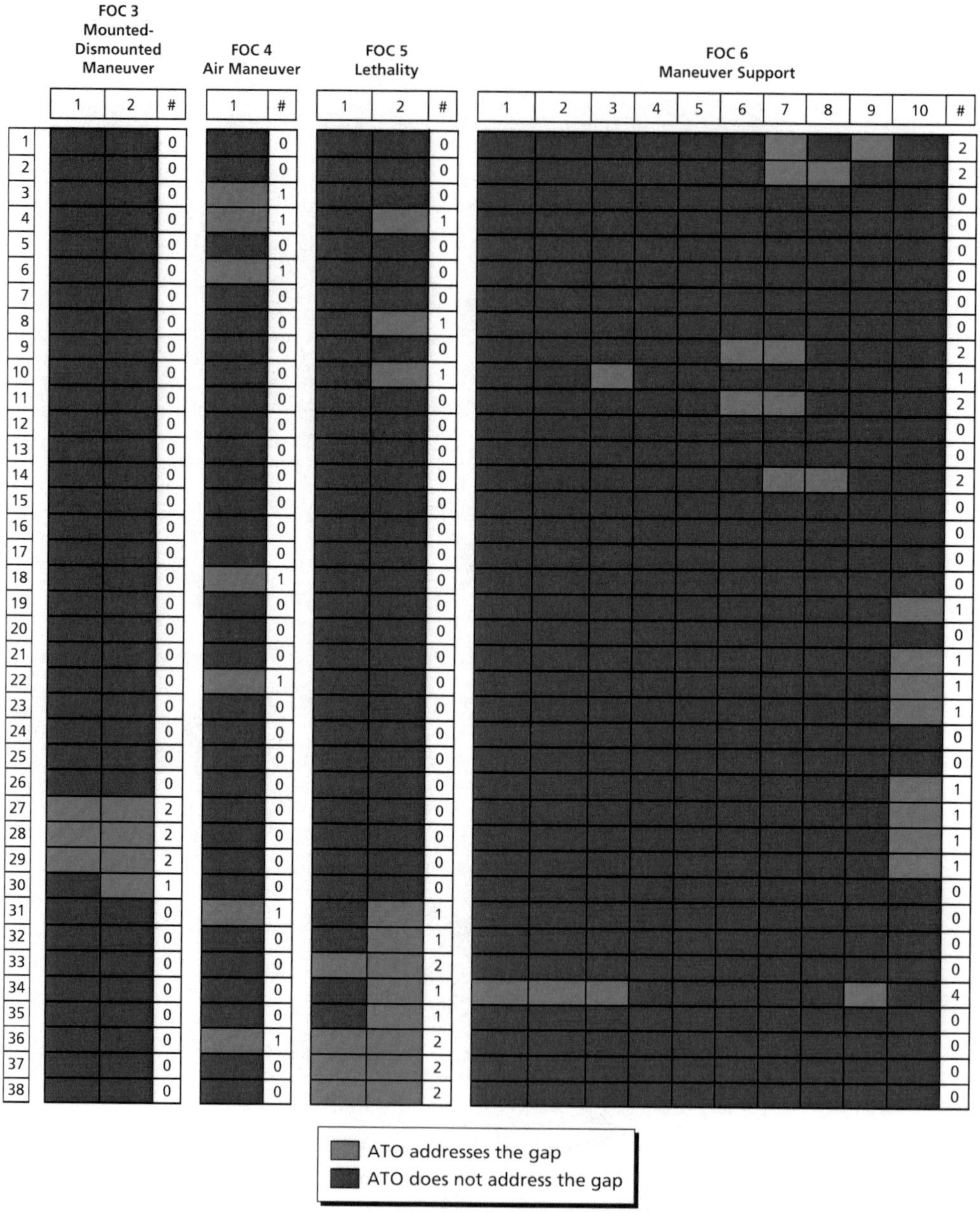

ATO number	FOC 3 Mounted-Dismounted Maneuver #	FOC 4 Air Maneuver #	FOC 5 Lethality #	FOC 6 Maneuver Support #
1	0	0	0	2
2	0	0	0	2
3	0	1	0	0
4	0	1	1	0
5	0	0	0	0
6	0	1	0	0
7	0	0	0	0
8	0	0	1	0
9	0	0	0	2
10	0	0	1	1
11	0	0	0	2
12	0	0	0	0
13	0	0	0	0
14	0	0	0	2
15	0	0	0	0
16	0	0	0	0
17	0	0	0	0
18	0	1	0	0
19	0	0	0	1
20	0	0	0	0
21	0	0	0	1
22	0	1	0	1
23	0	0	0	1
24	0	0	0	0
25	0	0	0	0
26	0	0	0	1
27	2	0	0	1
28	2	0	0	1
29	2	0	0	1
30	1	0	0	0
31	0	1	1	0
32	0	0	1	0
33	0	0	2	0
34	0	0	1	4
35	0	0	1	0
36	0	1	2	0
37	0	0	2	0
38	0	0	2	0

ATO addresses the gap
ATO does not address the gap

NOTES: The # column shows the total number of gaps in that FOC addressed by that ATO.
RAND *MG979-3.2*

Figure 3.3
ATOs 1–38 Gap-Space Coverage Matrix for FOC 7 Protection and FOC 8 Strategic Responsiveness and Deployability

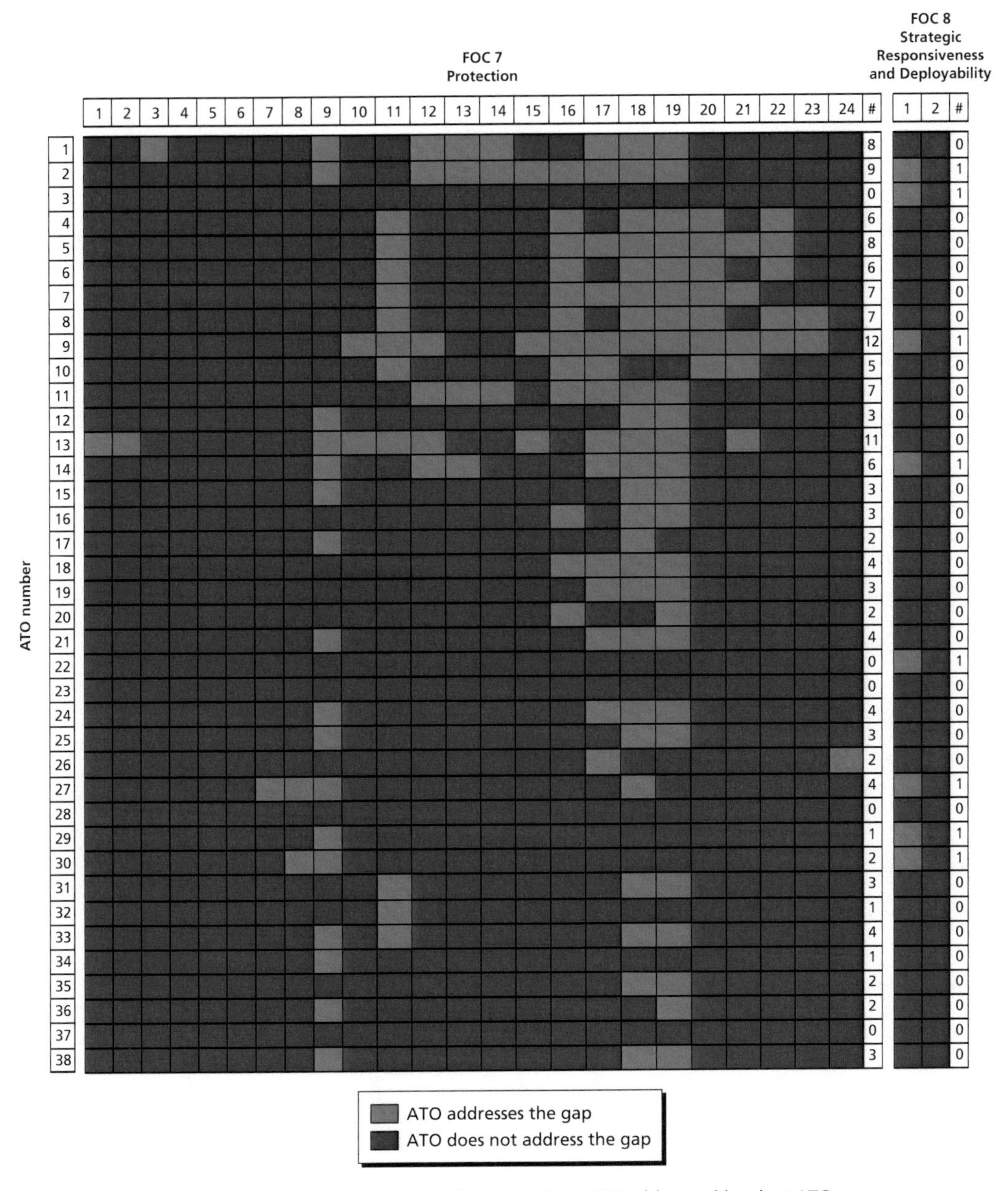

NOTE: The # column shows the total number of gaps in that FOC addressed by that ATO.

RAND *MG979-3.3*

Figure 3.4
ATOs 1–38 Gap-Space Coverage Matrix for FOC 9 Maneuver Sustainment; FOC 10 Training, Education, and Leadership; and FOC 11 Human Engineering

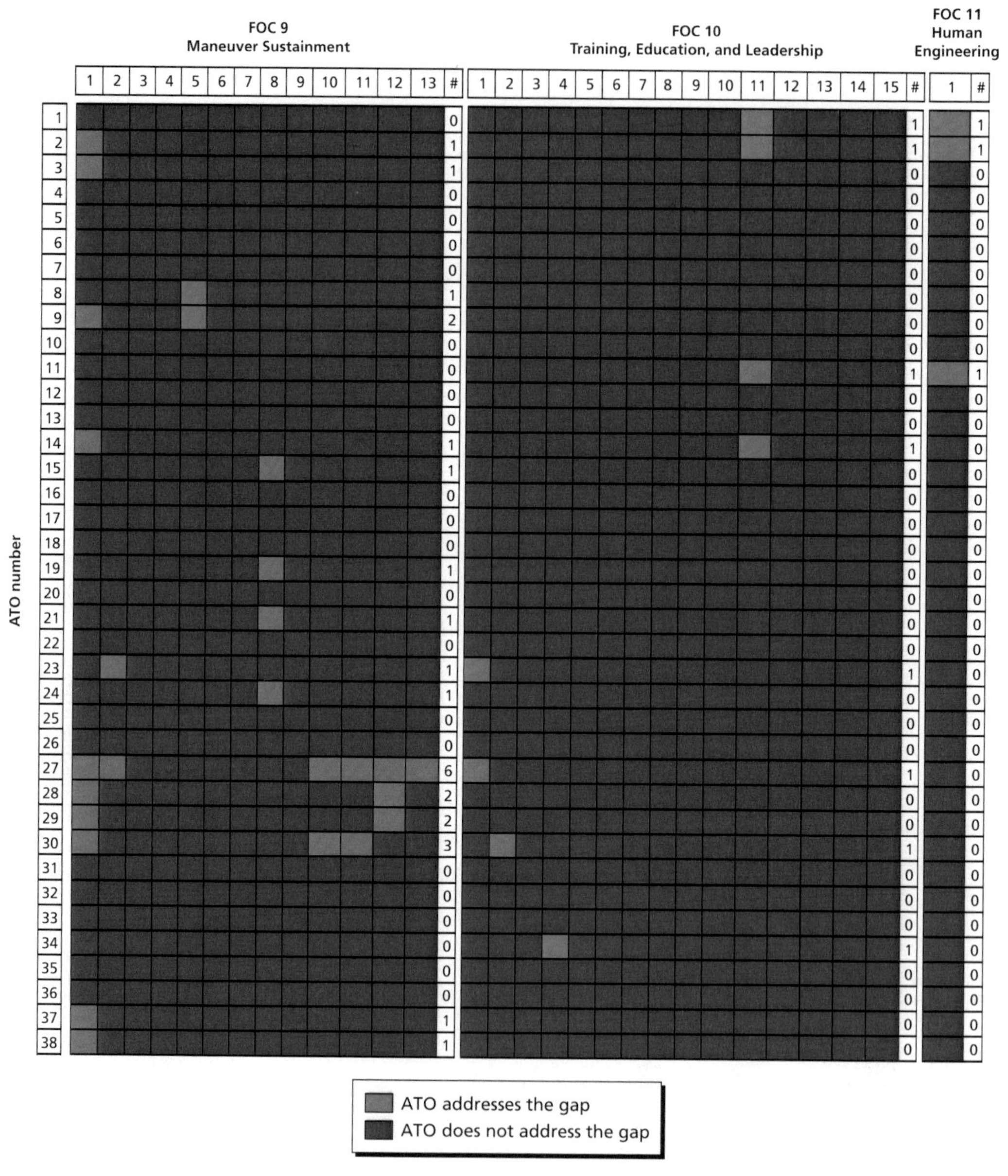

NOTE: The # column shows the total number of gaps in that FOC addressed by that ATO.
RAND *MG979-3.4*

Figure 3.5
ATOs 39–75 Gap-Space Coverage Matrix for FOC 1 Battle Command and FOC 2 Battlespace Awareness

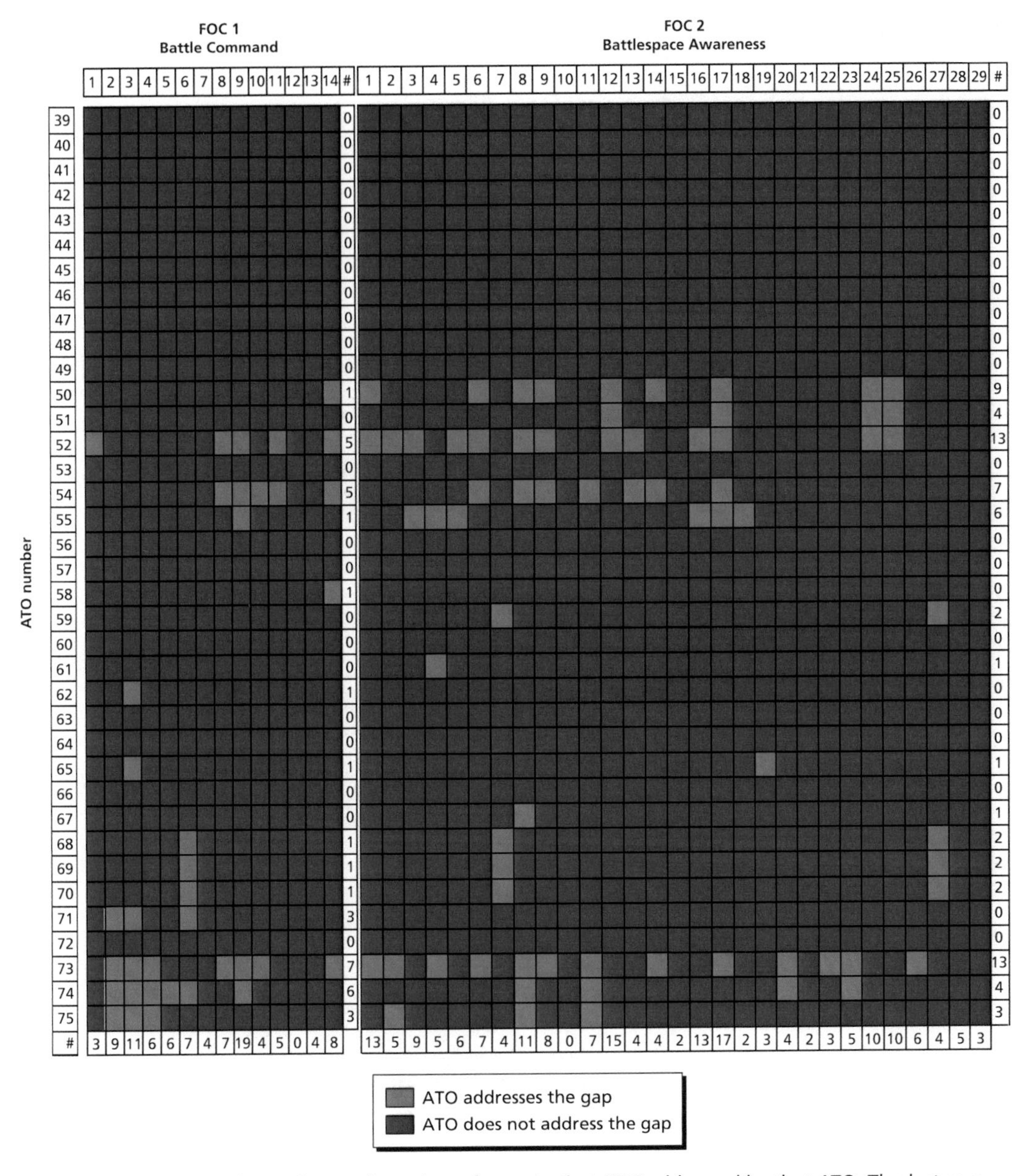

NOTES: The # column shows the total number of gaps in that FOC addressed by that ATO. The last row shows the number of contributing ATOs shown here and in Figure 3.1.

RAND *MG979-3.5*

Figure 3.6
ATOs 39–75 Gap-Space Coverage Matrix for FOC 3 Mounted-Dismounted Maneuver, FOC 4 Air Maneuver, FOC 5 Lethality, and FOC 6 Maneuver Support

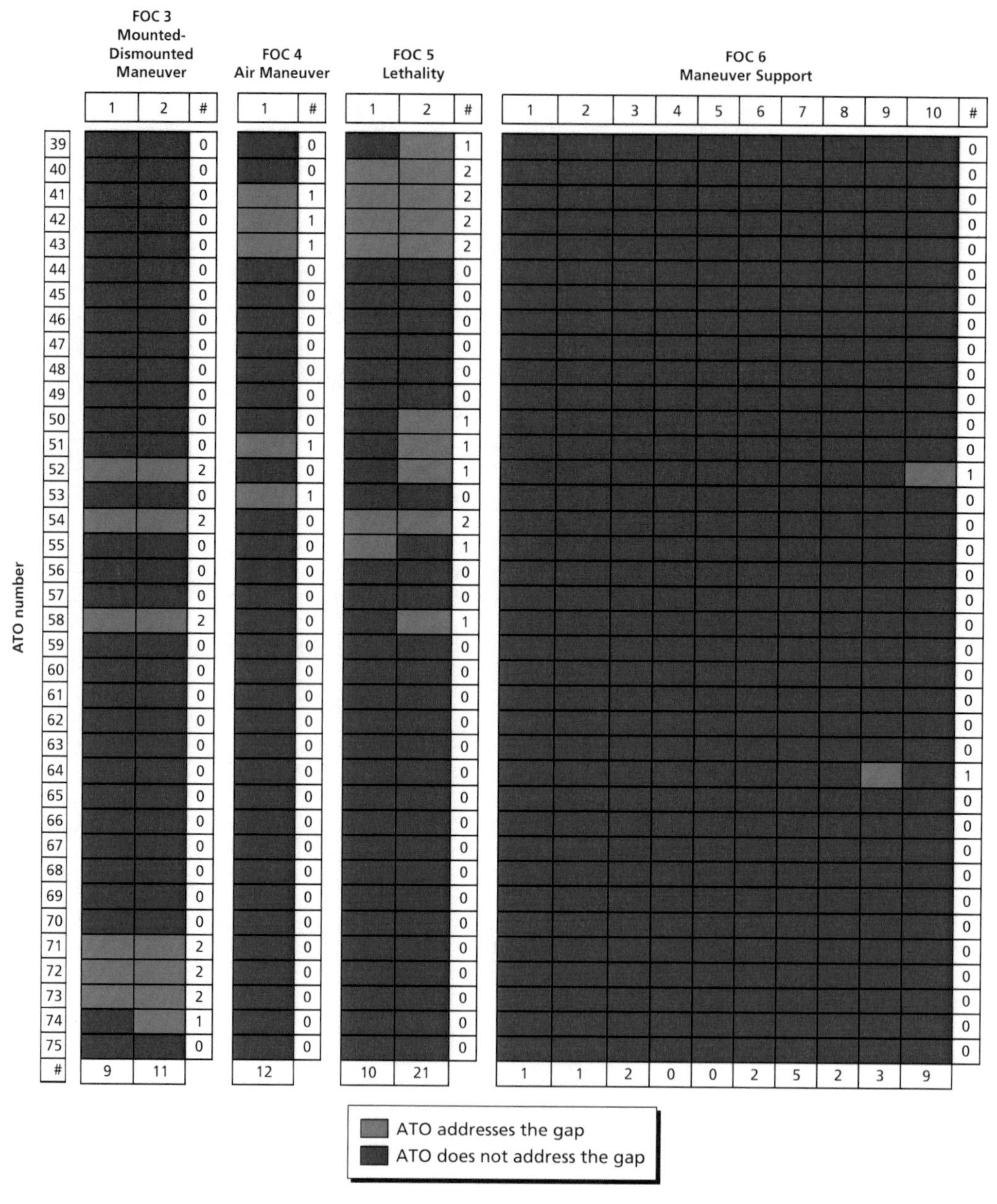

NOTES: The # column shows the total number of gaps in that FOC addressed by that ATO. The last row shows the number of contributing ATOs shown here and in Figure 3.2.

RAND *MG979-3.6*

Figure 3.7
ATOs 39–75 Gap-Space Coverage Matrix for FOC 7 Protection and FOC 8 Strategic Responsiveness and Deployability

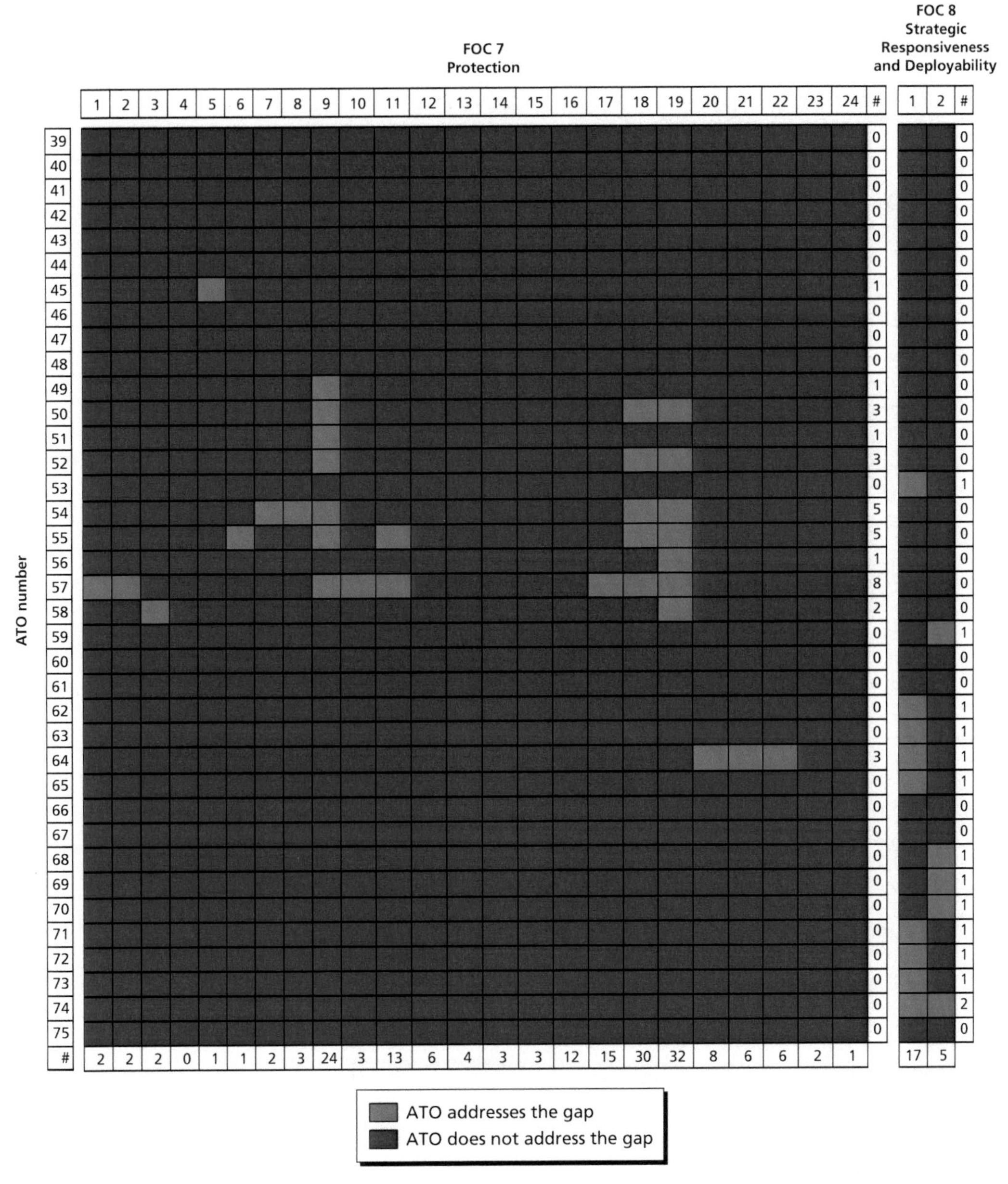

NOTES: The # column shows the total number of gaps in that FOC addressed by that ATO. The last row shows the number of contributing ATOs shown here and in Figure 3.3.

RAND *MG979-3.7*

Figure 3.8
ATOs 39–75 Gap-Space Coverage Matrix for FOC 9 Maneuver Sustainment; FOC 10 Training, Education, and Leadership; and FOC 11 Human Engineering

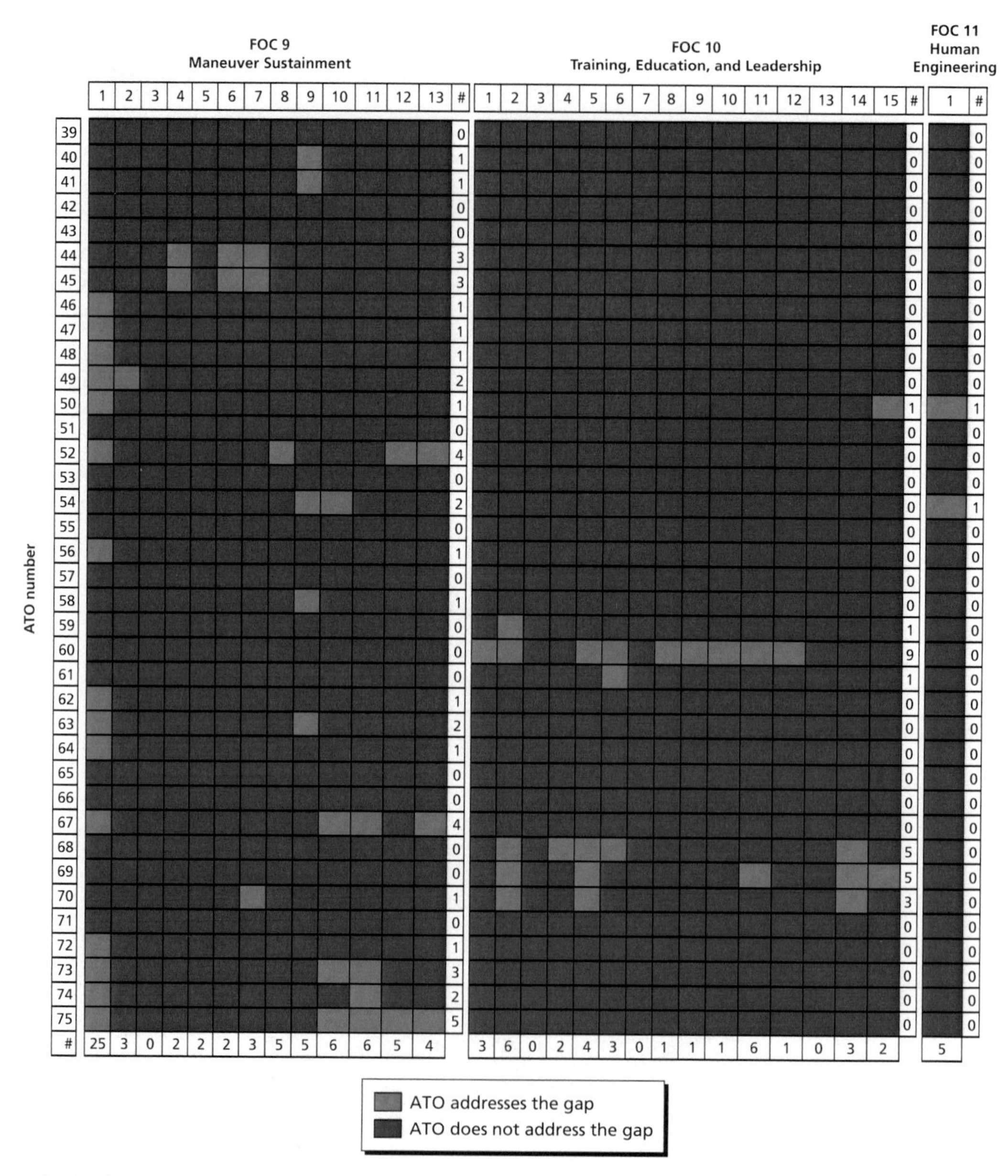

NOTES: The # column shows the total number of gaps in that FOC addressed by that ATO. The last row shows the number of contributing ATOs shown here and in Figure 3.4.

RAND *MG979-3.8*

Figure 3.9
Number of ATOs That Contribute to a Given Number of FOCs

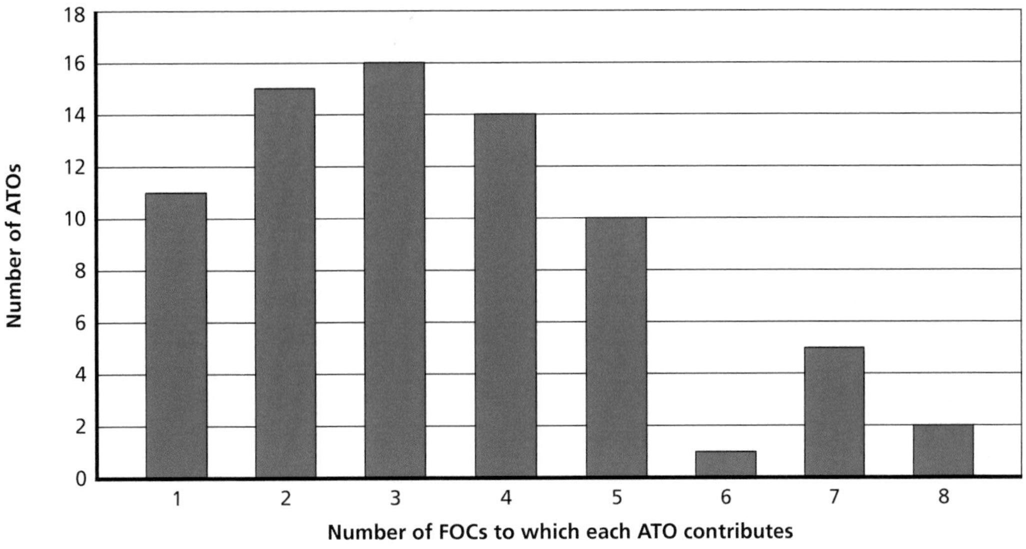

RAND *MG979-3.9*

Figure 3.13 combines the information in Figure 3.11 on gaps and Figure 3.12 on ATOs into an overall gap coverage matrix, disaggregated by FOC, situations, and categories. In the FOC boxes of Figure 3.13, the number in the upper left-hand corner is the number of gaps that apply to that FOC, and the number in the upper right-hand corner is the number of ATOs that address that set of gaps in that FOC. Similarly, for each element, the number in the upper left-hand corner is the number of gaps and the number in the upper right-hand corner is the number of ATOs, for that FOC, situation, and category. The range of colors indicates the level to which ATOs address the gaps, from dark green, indicating many more ATOs than gaps, to red, indicating many fewer ATOs than gaps. As in Figures 3.11 and 3.12, gray shaded boxes indicate no requirement.

The matrix of Figure 3.13 allows a focused analysis to quickly identify which gap areas are well supplied by ATOs, and which gap areas are being poorly addressed, down to the level of FOC categories and situations, which could be valuable data not only for ATO portfolio selection, as illustrated in this monograph, but also for the planning of future ATOs. Figure 3.13 is a demonstration of the analytical method, but the same method can be applied to real cases. For example, while the demonstration relies on the study team's judgments concerning assignment of gaps and ATOs to FOCs and their disaggregation according to situations and categories, a user can apply gap coverage to the real cases or use any other method desired to determine the contributions of individual ATOs to capability gaps.

Figure 3.10
Analytical Template

			Category			
FOC	1:Battle command (BC)	O	Command	Control	Computers	Communications
		E	Command	Control	Computers	Communications
		A	Command	Control	Computers	Communications
	2:Battlespace awareness (BSA)	O	LOS fire	B/NLOS fire	Force location	Hazards
		E	LOS fire	B/NLOS fire	Force location	Hazards
		A	LOS fire	B/NLOS fire	Force location	Hazards
	3:Mounted-dismounted maneuver (M-DM)	O	Forces	Mobility	Weapons	Equipment/supplies
		E	Forces	Mobility	Weapons	Equipment/supplies
		A	Forces	Mobility	Weapons	Equipment/supplies
	4:Air maneuver (AM)	O	Forces	Mobility	Weapons	Equipment/supplies
		E	Forces	Mobility	Weapons	Equipment/supplies
		A	Forces	Mobility	Weapons	Equipment/supplies
	5:Lethality (L)	O	LOS	BLOS	NLOS	
		E	LOS	BLOS	NLOS	
		A	LOS	BLOS	NLOS	
	6:Maneuver support (MSp)	O	Forces	Mobility	Weapons	Equipment/supplies
		E	Forces	Mobility	Weapons	Equipment/supplies
		A	Forces	Mobility	Weapons	Equipment/supplies
	7:Protection (P)	O	Disease	Injury	Asset protection	Information protection
		E	Disease	Injury	Asset protection	Information protection
		A	Disease	Injury	Asset protection	Information protection
	8:Strategic responsiveness and deployability (SRD)	O	Readiness	Transport	Delivery	
		E	Readiness	Transport	Delivery	
		A	Readiness	Transport	Delivery	
	9:Maneuver sustainment (MS)	O	Forces	Mobility	Weapons	Equipment/supplies
		E	Forces	Mobility	Weapons	Equipment/supplies
		A	Forces	Mobility	Weapons	Equipment/supplies
	10:Training, leadership, and education (TEL)	O	Doctrine	Equipment	People	Environment
		E	Doctrine	Equipment	People	Environment
		A	Doctrine	Equipment	People	Environment
	11:Human Engineering (HE)	O	Task/people	Systems	Interfaces	
		E	Task/people	Systems	Interfaces	
		A	Task/people	Systems	Interfaces	

NOTES: O = on the battlefield; E = en route or on the way to the battlefield; A = away from or off the battlefield; the black boxes signify that there is no fourth category for that FOC. These notes also apply to Figures 3.11–3.13.

RAND *MG979-3.10*

Figure 3.11
Distribution of Gaps by FOC, Situation, and Category

FOC			Category			
14	1:BC	O	14 Command	14 Control	5 Computers	12 Communications
		E	14 Command	14 Control	5 Computers	12 Communications
		A	11 Command	11 Control	4 Computers	9 Communications
29	2:BSA	O	17 LOS fire	16 B/NLOS fire	23 Force location	16 Hazards
		E	18 LOS fire	17 B/NLOS fire	24 Force location	17 Hazards
		A	20 LOS fire	19 B/NLOS fire	26 Force location	19 Hazards
2	3:M-DM	O	2 Forces	2 Mobility	2 Weapons	2 Equipment/supplies
		E	2 Forces	2 Mobility	2 Weapons	2 Equipment/supplies
		A	0 Forces	0 Mobility	0 Weapons	0 Equipment/supplies
1	4:AM	O	1 Forces	1 Mobility	1 Weapons	1 Equipment/supplies
		E	0 Forces	0 Mobility	0 Weapons	0 Equipment/supplies
		A	1 Forces	1 Mobility	1 Weapons	1 Equipment/supplies
2	5:L	O	2 LOS	1 BLOS	1 NLOS	
		E	0 LOS	0 BLOS	0 NLOS	
		A	0 LOS	0 BLOS	0 NLOS	
10	6:MSp	O	6 Forces	7 Mobility	3 Weapons	2 Equipment/supplies
		E	7 Forces	9 Mobility	3 Weapons	3 Equipment/supplies
		A	4 Forces	6 Mobility	1 Weapons	2 Equipment/supplies
24	7:P	O	6 Disease	21 Injury	12 Asset protection	3 Information protection
		E	5 Disease	20 Injury	13 Asset protection	3 Information protection
		A	4 Disease	10 Injury	6 Asset protection	3 Information protection
2	8:SRD	O	2 Readiness	1 Transport	1 Delivery	
		E	1 Readiness	0 Transport	0 Delivery	
		A	2 Readiness	1 Transport	1 Delivery	
13	9:MS	O	10 Forces	8 Mobility	2 Weapons	11 Equipment/supplies
		E	9 Forces	7 Mobility	2 Weapons	10 Equipment/supplies
		A	6 Forces	3 Mobility	0 Weapons	7 Equipment/supplies
15	10:TEL	O	4 Doctrine	11 Equipment	12 People	7 Environment
		E	4 Doctrine	8 Equipment	9 People	4 Environment
		A	4 Doctrine	7 Equipment	9 People	4 Environment
1	11:HE	O	0 Task/people	1 Systems	0 Interfaces	
		E	0 Task/people	1 Systems	0 Interfaces	
		A	0 Task/people	0 Systems	0 Interfaces	

NOTES: The numbers in the upper left-hand corners of the individual elements show how many gaps apply to that FOC, situation, and category. The elements shaded in gray indicate FOCs, situations, and categories for which there is no gap.

RAND *MG979-3.11*

Figure 3.12
Distribution of ATOs by FOC, Situation, and Category

FOC			Category							
1:BC	34	O	Command	20	Control	14	Computers	17	Communications	24
		E	Command	15	Control	11	Computers	16	Communications	18
		A	Command	17	Control	14	Computers	14	Communications	18
2:BSA	33	O	LOS fire	16	B/NLOS fire	13	Force location	20	Hazards	12
		E	LOS fire	14	B/NLOS fire	13	Force location	19	Hazards	11
		A	LOS fire	14	B/NLOS fire	12	Force location	19	Hazards	15
3:M-DM	11	O	Forces	7	Mobility	7	Weapons	7	Equipment/supplies	8
		E	Forces	6	Mobility	7	Weapons	7	Equipment/supplies	8
		A	Forces	4	Mobility	5	Weapons	5	Equipment/supplies	6
4:AM	12	O	Forces	2	Mobility	6	Weapons	10	Equipment/supplies	0
		E	Forces	1	Mobility	2	Weapons	2	Equipment/supplies	0
		A	Forces	1	Mobility	2	Weapons	2	Equipment/supplies	0
5:L	22	O	LOS	21	BLOS	16	NLOS	16		
		E	LOS	6	BLOS	6	NLOS	7		
		A	LOS	4	BLOS	5	NLOS	5		
6:MSp	17	O	Forces	13	Mobility	16	Weapons	7	Equipment/supplies	7
		E	Forces	13	Mobility	15	Weapons	7	Equipment/supplies	7
		A	Forces	12	Mobility	14	Weapons	5	Equipment/supplies	6
7:P	44	O	Disease	5	Injury	35	Asset protection	26	Information protection	4
		E	Disease	4	Injury	29	Asset protection	23	Information protection	4
		A	Disease	3	Injury	19	Asset protection	16	Information protection	4
8:SRD	21	O	Readiness	8	Transport	10	Delivery	10		
		E	Readiness	6	Transport	8	Delivery	8		
		A	Readiness	10	Transport	10	Delivery	10		
9:MS	38	O	Forces	21	Mobility	23	Weapons	18	Equipment/supplies	16
		E	Forces	19	Mobility	16	Weapons	10	Equipment/supplies	14
		A	Forces	16	Mobility	14	Weapons	7	Equipment/supplies	15
10:TEL	15	O	Doctrine	1	Equipment	9	People	2	Environment	8
		E	Doctrine	1	Equipment	9	People	2	Environment	8
		A	Doctrine	1	Equipment	9	People	1	Environment	8
11:HE	5	O	Task/people	5	Systems	1	Interfaces	1		
		E	Task/people	5	Systems	1	Interfaces	1		
		A	Task/people	2	Systems	0	Interfaces	0		

NOTES: The numbers in the upper right-hand corners of the individual elements show how many ATOs address that FOC, situation, and category. The elements shaded in gray indicate FOCs, situations, and categories for which there is no gap.

RAND *MG979-3.12*

Figure 3.13
Overall Gap Coverage Matrix

FOC	Gaps	ATOs	Situation	Category			
1:BC	14	34	O	14 Command 20	14 Control 14	5 Computers 17	12 Communications 24
			E	14 Command 15	14 Control 11	5 Computers 16	12 Communications 18
			A	11 Command 17	11 Control 14	4 Computers 14	9 Communications 18
2:BSA	29	33	O	17 LOS fire 16	16 B/NLOS fire 13	23 Force location 20	16 Hazards 12
			E	18 LOS fire 14	17 B/NLOS fire 13	24 Force location 19	17 Hazards 11
			A	20 LOS fire 14	19 B/NLOS fire 12	26 Force location 19	19 Hazards 15
3:M-DM	2	11	O	2 Forces 7	2 Mobility 7	2 Weapons 7	2 Equipment/supplies 8
			E	2 Forces 6	2 Mobility 7	2 Weapons 7	2 Equipment/supplies 8
			A	0 Forces 4	0 Mobility 5	0 Weapons 5	0 Equipment/supplies 6
4:AM	1	12	O	1 Forces 2	1 Mobility 6	1 Weapons 10	1 Equipment/supplies 0
			E	0 Forces 1	0 Mobility 2	0 Weapons 2	0 Equipment/supplies 0
			A	1 Forces 1	1 Mobility 2	1 Weapons 2	1 Equipment/supplies 0
5:L	2	22	O	2 LOS 21	1 BLOS 16	1 NLOS 16	
			E	0 LOS 6	0 BLOS 6	0 NLOS 7	
			A	0 LOS 4	0 BLOS 5	0 NLOS 5	
6:MSp	10	17	O	6 Forces 13	7 Mobility 16	3 Weapons 7	2 Equipment/supplies 7
			E	7 Forces 13	9 Mobility 15	3 Weapons 7	3 Equipment/supplies 7
			A	4 Forces 12	6 Mobility 14	1 Weapons 5	2 Equipment/supplies 6
7:P	24	44	O	6 Disease 5	21 Injury 35	12 Asset protection 26	3 Information protection 4
			E	5 Disease 4	20 Injury 29	13 Asset protection 23	3 Information protection 4
			A	4 Disease 3	10 Injury 19	6 Asset protection 16	3 Information protection 4
8:SRD	2	21	O	2 Readiness 8	1 Transport 10	1 Delivery 10	
			E	1 Readiness 6	0 Transport 8	0 Delivery 8	
			A	2 Readiness 10	1 Transport 10	1 Delivery 10	
9:MS	13	38	O	10 Forces 21	8 Mobility 23	2 Weapons 18	11 Equipment/supplies 16
			E	9 Forces 19	7 Mobility 16	2 Weapons 10	10 Equipment/supplies 14
			A	6 Forces 16	3 Mobility 14	0 Weapons 7	7 Equipment/supplies 15
10:TEL	15	15	O	4 Doctrine 1	11 Equipment 9	12 People 2	7 Environment 8
			E	4 Doctrine 1	8 Equipment 9	9 People 2	4 Environment 8
			A	4 Doctrine 1	7 Equipment 9	9 People 1	4 Environment 8
11:HE	1	5	O	0 Task/people 5	1 Systems 1	0 Interfaces 1	
			E	0 Task/people 5	1 Systems 1	0 Interfaces 1	
			A	0 Task/people 2	0 Systems 0	0 Interfaces 0	

More ATOs than gaps — Fewer ATOs than gaps

NOTES: The numbers in the upper left-hand corners of the individual elements show how many gaps apply to that FOC, situation, and category. The numbers in the upper right-hand corners of the individual elements show how many ATOs address that FOC, situation, and category. The elements shaded in gray indicate FOCs, situations, and categories for which there is no gap.

RAND *MG979-3.13*

ATO Expected Values

We can use the gap-space coverage matrix of Figures 3.1 through 3.8, together with the method described in Chapter Two, to make expected value estimates for the ATOs. To do this, we need to perform, for each matrix element in Figures 3.1 through 3.8, the following analyses:

- Determine the situations and categories to which the gap applies.
- Determine which situations and categories the ATO addresses.
- Estimate the coverage score of the ATO for that gap as the fraction of the situations times the fraction of the categories to which the gap applies that the ATO addresses.
- Multiply by one-half (the assumed technical potential of each ATO, as described in Chapter Two) to convert the coverage score to the estimated expected value of the ATO for that gap.

The study team performed the analyses described in the previous paragraph for the entire gap coverage matrix, with the results shown in Figures A.1 through A.8 of Appendix A. We then obtained expected value estimates for the ATOs for each FOC by adding the contributions from each individual gap within that FOC and dividing by the number of gaps, according to the procedure described in Chapter Two. These ATO expected values are shown in the ATO-FOC matrix of Figures 3.14 and 3.15.[4] We note that because of the likelihood that some ATOs will fail to be successfully completed, it is desirable for the FOC EVs (i.e., the sum of the appropriate columns in Figures 3.14 and 3.15) to be greater than 100 percent, as most are. However, one of the FOCs has the total expected value (TEV) from all ATOs of less then 100 percent. We will analyze the effect of uncertainty on meeting the FOC requirements in Chapter Four.

Thus far, we have used a simplifying assumption in this chapter, namely, that ATOs' contributions to filling capability gaps within the same FOC are substitutable and additive. In other words, contributions enough to fill the same gap twice are as good as contributions that fill two gaps fully for the same FOC. On one hand, this assumption may be acceptable, as filling the same gap twice can in practice mean that the ATOs will not only make the capability gap disappear, but will also result in that capability being performed better than expected, which is welcome. On the other hand, if the user feels that the assumption is unsound, the same method can still be

[4] We repeat that the purpose of the EV estimates presented in Appendix A, and in Figures 3.14 and 3.15, is to demonstrate our method, not to make decisions on ATOs, since we do not have adequate data to determine the technical potential for each ATO and have arbitrarily assigned the same factor of one-half to all ATOs. An Excel spreadsheet that identifies the gaps and shows the study team's assignments of gaps and ATOs to situations and FOC categories is available with the approval of the sponsor of this study. If interested, please see the RAND contact information in the Preface of this monograph.

Figure 3.14
Expected Values for ATOs 1–38 (percent)

ATO number	FOC										
	1	2	3	4	5	6	7	8	9	10	11
1	0.00	0.00	0.00	0.00	0.00	10.00	14.93	0.00	0.00	1.67	50.00
2	3.57	3.02	0.00	0.00	0.00	10.00	17.01	8.33	1.28	1.67	50.00
3	0.00	0.00	0.00	12.50	0.00	0.00	0.00	12.50	1.28	0.00	0.00
4	0.00	0.00	0.00	12.50	8.33	0.00	6.60	0.00	0.00	0.00	0.00
5	0.00	0.00	0.00	0.00	0.00	0.00	15.28	0.00	0.00	0.00	0.00
6	0.00	0.00	0.00	18.75	0.00	0.00	6.60	0.00	0.00	0.00	0.00
7	0.00	0.00	0.00	0.00	0.00	0.00	13.89	0.00	0.00	0.00	0.00
8	0.00	0.00	0.00	0.00	8.33	0.00	7.64	0.00	0.96	0.00	0.00
9	0.00	0.00	0.00	0.00	0.00	5.00	24.31	25.00	5.13	0.00	0.00
10	0.79	0.00	0.00	0.00	8.33	1.67	10.42	0.00	0.00	0.00	0.00
11	0.79	0.00	0.00	0.00	0.00	10.00	13.89	0.00	0.00	1.11	50.00
12	0.00	0.00	0.00	0.00	0.00	0.00	1.91	0.00	0.00	0.00	0.00
13	0.00	0.00	0.00	0.00	0.00	0.00	17.71	0.00	0.00	0.00	0.00
14	0.00	3.02	0.00	0.00	0.00	10.00	10.76	16.67	1.28	1.67	0.00
15	2.38	5.75	0.00	0.00	0.00	0.00	4.51	0.00	2.56	0.00	0.00
16	3.57	1.72	0.00	0.00	0.00	0.00	5.56	0.00	0.00	0.00	0.00
17	4.31	8.29	0.00	0.00	0.00	0.00	1.74	0.00	0.00	0.00	0.00
18	3.57	10.34	0.00	25.00	0.00	0.00	7.64	0.00	0.00	0.00	0.00
19	3.57	7.33	0.00	0.00	0.00	5.00	5.56	0.00	1.28	0.00	0.00
20	1.19	3.74	0.00	0.00	0.00	0.00	3.13	0.00	0.00	0.00	0.00
21	3.57	15.52	0.00	0.00	0.00	5.00	6.60	0.00	1.28	0.00	0.00
22	3.57	1.29	0.00	12.50	0.00	2.50	0.00	25.00	0.00	0.00	0.00
23	3.57	7.76	0.00	0.00	0.00	2.50	0.00	0.00	3.85	3.33	0.00
24	3.57	4.31	0.00	0.00	0.00	0.00	6.60	0.00	2.56	0.00	0.00
25	2.38	4.24	0.00	0.00	0.00	0.00	2.78	0.00	0.00	0.00	0.00
26	3.57	3.45	0.00	0.00	0.00	5.00	2.78	0.00	0.00	0.00	0.00
27	28.57	15.52	50.00	0.00	0.00	5.00	6.94	16.67	23.08	3.33	0.00
28	12.50	10.34	50.00	0.00	0.00	5.00	0.00	0.00	7.69	0.00	0.00
29	36.31	8.62	50.00	0.00	0.00	5.00	1.04	16.67	7.69	0.00	0.00
30	28.57	2.16	6.25	0.00	0.00	0.00	3.47	25.00	6.09	3.33	0.00
31	0.00	0.00	0.00	6.25	25.00	0.00	1.39	0.00	0.00	0.00	0.00
32	0.00	0.00	0.00	0.00	16.67	0.00	1.04	0.00	0.00	0.00	0.00
33	0.00	0.00	0.00	0.00	50.00	0.00	3.82	0.00	0.00	0.00	0.00
34	0.00	0.00	0.00	0.00	25.00	17.50	1.04	0.00	0.00	1.11	0.00
35	0.00	0.00	0.00	0.00	8.33	0.00	1.74	0.00	0.00	0.00	0.00
36	0.40	0.00	0.00	6.25	50.00	0.00	2.08	0.00	0.00	0.00	0.00
37	0.00	0.00	0.00	0.00	50.00	0.00	0.00	0.00	2.56	0.00	0.00
38	0.00	0.00	0.00	0.00	50.00	0.00	2.78	0.00	2.56	0.00	0.00

RAND *MG979-3.14*

Figure 3.15
Expected Values for ATOs 39–75 (percent)

ATO number	FOC										
	1	2	3	4	5	6	7	8	9	10	11
39	0.00	0.00	0.00	0.00	25.00	0.00	0.00	0.00	0.00	0.00	0.00
40	0.00	0.00	0.00	0.00	50.00	0.00	0.00	0.00	1.44	0.00	0.00
41	0.00	0.00	0.00	6.25	50.00	0.00	0.00	0.00	0.48	0.00	0.00
42	0.00	0.00	0.00	6.25	50.00	0.00	0.00	0.00	0.00	0.00	0.00
43	0.00	0.00	0.00	12.50	50.00	0.00	0.00	0.00	0.00	0.00	0.00
44	0.00	0.00	0.00	0.00	0.00	0.00	0.00	0.00	2.99	0.00	0.00
45	0.00	0.00	0.00	0.00	0.00	0.00	0.69	0.00	1.50	0.00	0.00
46	0.00	0.00	0.00	0.00	0.00	0.00	0.00	0.00	1.28	0.00	0.00
47	0.00	0.00	0.00	0.00	0.00	0.00	0.00	0.00	1.28	0.00	0.00
48	0.00	0.00	0.00	0.00	0.00	0.00	0.00	0.00	1.28	0.00	0.00
49	0.00	0.00	0.00	0.00	0.00	0.00	2.08	0.00	6.41	0.00	0.00
50	4.37	10.78	0.00	0.00	41.67	0.00	5.56	0.00	2.56	1.21	50.00
51	0.00	5.17	0.00	12.50	25.00	0.00	1.04	0.00	0.00	0.00	0.00
52	11.90	12.07	50.00	0.00	25.00	5.00	4.51	0.00	12.82	0.00	0.00
53	0.00	0.00	0.00	12.50	0.00	0.00	0.00	12.50	0.00	0.00	0.00
54	13.49	9.20	50.00	0.00	50.00	0.00	3.70	0.00	6.41	0.00	50.00
55	0.85	2.48	0.00	0.00	25.00	0.00	3.13	0.00	0.00	0.00	0.00
56	0.00	0.00	0.00	0.00	0.00	0.00	1.04	0.00	2.56	0.00	0.00
57	0.00	0.00	0.00	0.00	0.00	0.00	8.33	0.00	0.00	0.00	0.00
58	0.79	0.00	18.75	0.00	25.00	0.00	1.56	0.00	1.44	0.00	0.00
59	0.00	3.45	0.00	0.00	0.00	0.00	0.00	8.33	0.00	1.11	0.00
60	0.00	0.00	0.00	0.00	0.00	0.00	0.00	0.00	0.00	20.56	0.00
61	0.00	1.72	0.00	0.00	0.00	0.00	0.00	0.00	0.00	3.33	0.00
62	0.79	0.00	0.00	0.00	0.00	0.00	0.00	8.33	2.56	0.00	0.00
63	0.00	0.00	0.00	0.00	0.00	0.00	0.00	16.67	3.21	0.00	0.00
64	0.00	0.00	0.00	0.00	0.00	1.67	6.25	16.67	2.56	0.00	0.00
65	0.79	0.43	0.00	0.00	0.00	0.00	0.00	8.33	0.00	0.00	0.00
66	0.00	0.00	0.00	0.00	0.00	0.00	0.00	0.00	0.00	0.00	0.00
67	0.00	0.43	0.00	0.00	0.00	0.00	0.00	0.00	11.22	0.00	0.00
68	1.19	3.45	0.00	0.00	0.00	0.00	0.00	8.33	0.00	5.56	0.00
69	1.19	3.45	0.00	0.00	0.00	0.00	0.00	8.33	0.00	4.54	0.00
70	0.40	0.86	0.00	0.00	0.00	0.00	0.00	8.33	0.85	3.06	0.00
71	5.56	0.00	12.50	0.00	0.00	0.00	0.00	5.56	0.00	0.00	0.00
72	0.00	0.00	37.50	0.00	0.00	0.00	0.00	16.67	2.56	0.00	0.00
73	25.00	22.41	25.00	0.00	0.00	0.00	0.00	8.33	8.33	0.00	0.00
74	21.43	6.90	25.00	0.00	0.00	0.00	0.00	33.33	7.69	0.00	0.00
75	6.55	2.59	0.00	0.00	0.00	0.00	0.00	0.00	12.18	0.00	0.00

NOTE: The EVs for ATO 66 are nonzero beyond the first two digits but rounded to zero.

RAND *MG979-3.15*

used by splitting the FOC into two or more sub-FOCs and treating them as independent FOCs. This shows the flexibility of an LPM to accommodate special situations.

Importance of Individual ATOs in Filling Gaps

The magnitude of the expected values of ATOs for FOCs shown in Figures 3.14 and 3.15 provides one indication of the relative importance of ATOs. However, certain ATOs and groups of ATOs provide a significant portion, and in some cases, all, of the expected value of contributions to FOCs and their capability gaps. Moreover, some ATOs provide additional capabilities to gaps and FOCs that are already well covered. In such cases, the possibility exists to reduce funding for these ATOs and use these funds for new ATOs that address gaps that are either uncovered or poorly covered by the existing ATOs.

Below we first list ATOs that are important because they cover gaps that are covered by no other ATOs or by only one other ATO. These ATOs would likely be included in portfolios that require all gaps be covered, rather than just meeting individual FOC EV requirements. When considering the possibility of ATO failure, these ATOs take on special significance, because they represent cases in which failure of just one ATO leads to uncovered gaps or gaps that are just one more failed ATO from being uncovered.

We also list a second set of important ATOs—those that contribute a significant fraction (i.e., greater than 10 percent) of the expected value of individual FOCs. The intersection of these two groups of ATOs represents a third group of especially important ATOs—those that cover gaps that are otherwise uncovered or are covered by only one other ATO *and* also contribute a significant fraction to meeting individual FOC gap requirements.

These lists show ATOs that are of special importance in meeting FOC gap and EV requirements. We are also interested in which ATOs play a lesser role in meeting these requirements. The fourth list shows those ATOs whose average contribution to FOC expected values is less than 0.5 percent. Except for ATOs 8, 25, 44, 45, and 57, which are also in the first list and are of special importance because they cover gaps that are covered by no (or only one) other ATO, the ATOs of the fourth list may be of lesser importance to meeting FOC gap requirements and thus can be considered candidates for reduced funding to allow additional funds to be applied to meeting currently unmet FOC gap requirements. Alternatively, the appearance of certain ATOs in the fourth list (e.g., ATOs 46–48 on vaccines) may indicate that new capability gaps are required to address soldier needs.

ATOs That Cover Gaps That Are Covered by No Other ATOs or by Only One Other ATO[5]

- ATO 1: network electronic warfare ATO [force protection]
- ATO 2: mine and IED detection ATO [force protection]
- ATO 8: pulse power for the FCS ATO [force protection]
- ATO 9: vehicle armor technology ATO [force protection]
- ATO 10: solid-state laser technology ATO [force protection]
- ATO 11: countermine and IED neutralization ATO [force protection]
- ATO 13: modular protective systems for Future Force assets ATO [force protection]
- ATO 14: wide area airborne minefield detection ATO [force protection]
- ATO 15: third-generation IR technologies ATO [ISR]
- ATO 17: suite of sense-through-the-wall systems ATO [ISR]
- ATO 21: class-II UAV EO payloads ATO [ISR]
- ATO 24: low-cost, high-resolution IR focal plane arrays ATO [ISR]
- ATO 25: human infrastructure detection and exploitation ATO [ISR]
- ATO 26: tactical wireless network assurance ATO [C4]
- ATO 27: networked enabled command and control ATO [C4]
- ATO 34: nonlethal payloads for personnel suppression ATO [lethality]
- ATO 44: Automated Critical Care Life Support System ATO [medical]
- ATO 45: fluid resuscitation technology to reduce injury and loss of life on the battlefield ATO [medical]
- ATO 50: robotics collaboration ATO and near autonomous unmanned systems ATO [unmanned systems]
- ATO 54: Future Force Warrior ATD [soldier systems]
- ATO 55: soldier mobility vision systems ATO [soldier systems]
- ATO 57: soldier protection technologies ATO [soldier systems]
- ATO 58: mounted/dismounted soldier power ATO [soldier systems]
- ATO 60: leader adaptability ATO [soldier systems]
- ATO 68: learning with adaptive simulation and training ATO [advanced simulation]
- ATO 69: scaleable embedded training and mission rehearsal ATO [advanced simulation]

ATOs and ACTD/JCTDs That Contribute at Least 10 Percent of the Expected Value of an Individual FOC

- ATO 1: network electronic warfare ATO [force protection]
- ATO 2: mine and IED detection ATO [force protection]
- ATO 6: extended-area protection and survivability ATO [force protection]

[5] ATO technology types are shown in brackets in this list and those lists below.

- ATO 11: countermine and IED neutralization ATO [force protection]
- ATO 18: multimission radar ATO [ISR]
- ATO 27: networked enabled command and control ATO [C4]
- ATO 28: tactical mobile networks ATO [C4]
- ATO 29: tactical network and communications antennas ATO [C4]
- ATO 30: battlespace terrain reasoning and awareness—battle command ATO [C4]
- ATO 34: nonlethal payloads for personnel suppression ATO [lethality]
- ATO 50: robotics collaboration ATO and near autonomous unmanned systems ATO [unmanned systems]
- ATO 52: Army/DARPA enabling technologies for the FCS ATO [unmanned systems]
- ATO 54: Future Force Warrior ATD [soldier systems]
- ATO 60: leader adaptability ATO [soldier systems]
- ATO 73: adaptive joint C4ISR node ACTD [ACTD/JCTD]
- ATO 74: theater effects–based operations ACTD [ACTD/JCTD]

ATOs That Are Included in Both Lists Above

- ATO 1: network electronic warfare ATO [force protection]
- ATO 2: mine and IED detection ATO [force protection]
- ATO 11: countermine and IED neutralization ATO [force protection]
- ATO 27: networked enabled command and control ATO [C4]
- ATO 34: nonlethal payloads for personnel suppression ATO [lethality]
- ATO 50: robotics collaboration ATO and near autonomous unmanned systems ATO [unmanned systems]
- ATO 54: Future Force Warrior ATD [soldier systems]
- ATO 60: leader adaptability ATO [soldier systems]

ATOs That on Average Contribute Less Than 0.5 Percent of the Expected Value of Individual FOCs

- ATO 5: Passive Infrared Cueing System ATO [force protection]
- ATO 7: dissemination of advanced obscurants ATO [force protection]
- ATO 8: pulse power for the FCS ATO [force protection]
- ATO 12: vision protection ATO [force protection]
- ATO 16: Distributed Aperture System ATO [ISR]
- ATO 20: distributed imaging radar technology for continuous battlefield imagery ATO [ISR]
- ATO 25: human infrastructure detection and exploitation ATO [ISR]
- ATO 32: Mounted Combat System and Abrams Ammunition System technologies ATO [lethality]

- ATO 35: electromagnetic gun technology maturation and demonstration ATO [lethality]
- ATO 39: hardened combined effects Penetrator warheads ATO [lethality]
- ATO 44: Automated Critical Care Life Support System ATO [medical]
- ATO 45: fluid resuscitation technology to reduce injury and loss of life on the battlefield ATO [medical]
- ATO 46: vaccines and drugs to prevent and treat malaria ATO [medical]
- ATO 47: vaccines to prevent diarrhea ATO [medical]
- ATO 48: vaccine for the prevention of military HIV infection ATO [medical]
- ATO 49: biomedical enablers of operational health and performance ATO [medical]
- ATO 56: nutritionally optimized first strike ration ATO [soldier systems]
- ATO 57: soldier protection technologies ATO [soldier systems]
- ATO 62: precision airdrop–medium ATO [logistics]
- ATO 65: joint rapid airfield construction ATO [logistics]
- ATO 66: JP-8 reformer for alternate fuel sources ATO [logistics].

CHAPTER FOUR

Applications of the Full Method

This study takes a two-step approach to analyze the ATO portfolio. In Chapter Three, we used the gap-space method to identify broadly where the Army may encounter problems meeting requirements with existing S&T projects. This examination was performed under the assumption that all ongoing ATOs would continue to be funded and completed successfully. Therefore, this examination can be used as a map of supply and demand, showing where requirements are not met even under the most optimistic assumption that none of the ATOs fails. In this chapter, we refine this map given that some ATOs in the portfolio will, in reality, fail and not lead to a fielded system. For the refinements and other applications described in the bullets below, we employ the full method developed during this study. In addition to the gap-space method for estimating expected values, the full method includes the Delphi method for estimating marginal implementation costs; the LPM, with the assumption that the success of ATOs is certain; and the simulation, with the assumption that the success of ATOs is uncertain. We arrange the applications into a chronological series of decisions that Army managers will encounter in selecting and managing the ATOs over the course of the program, namely the following:

- Determine to what extent the ongoing ATOs can meet capability gaps cost-effectively, as well as the remaining gaps that can be filled more cheaply with new ATOs. This determination is described in the section "Establish Realistic FOC Requirements for Existing ATOs."
- Determine a cost-effective total remaining lifecycle budget for the ongoing ATOs; any amount beyond this should not be used to fund ongoing ATOs, but rather to fund new ATOs and their systems, as developing new ATOs is a more cost-effective use of these funds. This determination is described in the section "Determine Total Remaining Lifecycle Budget for Existing ATOs."
- Optimally divide total remaining lifecycle funds between total remaining S&T budget and total marginal implementation budget for the ongoing ATOs so that the chance to meet all requirements is maximized. This division is described in the section "Determine the Optimal S&T Budget."

- Decide which ongoing ATOs should be kept for continued funding when the total remaining S&T budget cannot support all ongoing ATOs. The decision process is described in the section "Select ATOs for Continued Funding."
- Determine the impact on the likelihood to meet all requirements (feasible percentage) if the total remaining S&T budget is below optimal. This determination is described in the section "Impact of Suboptimal Total Remaining S&T Budget."
- Determine the remaining S&T cost and the remaining lifecycle cost needed to fill each capability gap. This determination is described in the section "Optimal Distribution of Funds Among FOC Gaps."

Establish Realistic FOC Requirements for Existing ATOs

FOC gaps and S&T projects can be treated as demand and supply of capabilities, respectively. Therefore, the Army first identifies the FOC gap requirements. Then, the S&T community develops S&T projects so their end products can be used to meet the requirements.[1] Figure 4.1, based on the EV estimates shown in Figures 3.14 and 3.15, shows the maximum potential of the 75 ongoing ATOs in meeting the 11 categories of FOC gaps. This is the largest possible contribution of ATOs to meeting FOC

Figure 4.1
Maximum Potential of 75 ATOs to Meet 11 Unmodified FOC Gaps

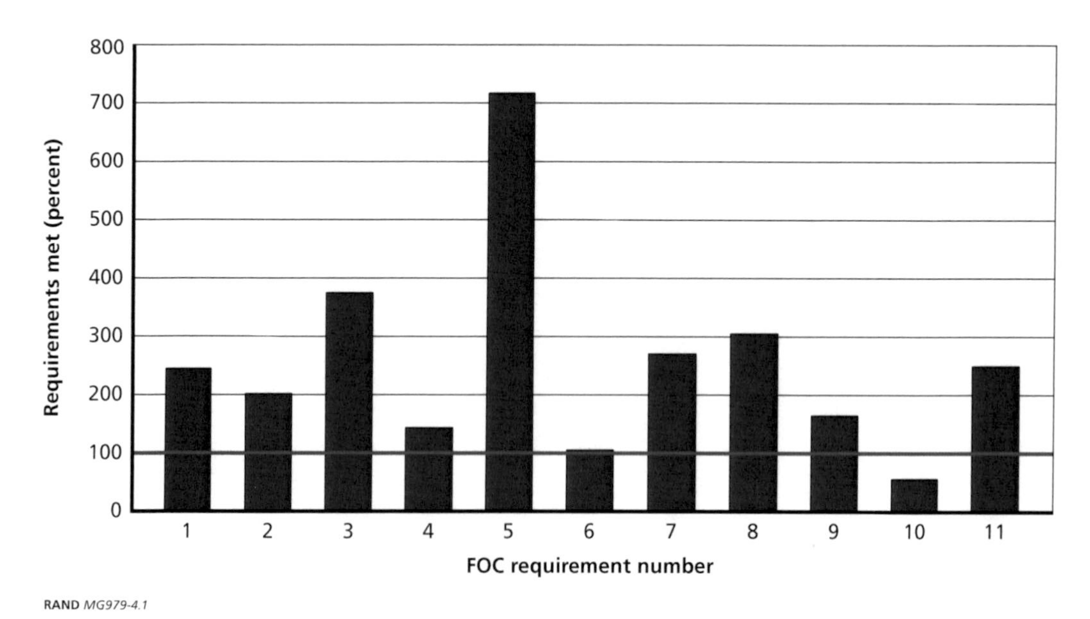

RAND *MG979-4.1*

---

1 The balance of demand and supply is an iterative process. In setting the requirements, a planner would study the S&T program and the technology advances so as to see whether the requirements will possibly be filled.

gaps, because all 75 ATOs are assumed to have a 100-percent probability of success. In other words, if all ATOs are certain to be successfully completed to meet their goals for system performance and cost, all 11 FOC requirements will be met, except FOC 10, where only 57 percent can possibly be met. Clearly, if planners were to ask whether these existing ATOs can meet all 11 FOC requirements, the answer is no; since even with perfect ATO success, FOC 10 requirement will not be met.

Since not all ATOs will succeed in reality, the obvious next question is what other FOCs are at risk if some of the ATOs that contribute to their filling gaps are unsuccessful. If we can determine the at-risk FOCs, as well as those FOCs most overmet by the ATO portfolio, a better allocation of funds to existing and new ATO resources can be made—eliminating some existing ATOs contributing to the overmet FOCs and initiating new ATOs specifically tailored to better cover the at-risk FOCs. To determine the at-risk and overmet FOCs, we return to the concept of feasible percentage that was introduced at the end of Chapter Two.

Using the simulation technique described in Chapter Two, we found that if all ATOs had a 90-percent probability of success, there was only a 16.3-percent chance (feasible percentage) of meeting the best-case FOC-scenario—all FOC requirements met, except for FOC 10, where only 57 percent of its requirements are met. This is shown in row 1 of Table 4.1. Assessing the impact on feasible percentage of removing (discontinuing the funding of) the largest EV-contributing ATO for each FOC provides insight into which other FOCs are at risk. Row 2 of Table 4.1 shows that removing the largest EV-contributing ATO for each FOC results in a gap in FOC 6 (it is only 88 percent covered) and the FOC 10 coverage also declines to 36 percent.

Interestingly, these somewhat-reduced requirement numbers can be interpreted from a different perspective. Since Table 4.1 shows a "worst-case" scenario that removes the largest EV-contributing ATO to each FOC, these reduced requirements will also be met if a lesser EV-contributing ATO to each FOC were removed or failed.[2] We

Table 4.1
Cost-Effective Trade: A Small Reduction in Gap Requirements for a Large Increase in Feasible Percentage

Number of highest contributers removed	Percentage feasible	FOC requirements (%)										
		1	2	3	4	5	6	7	8	9	10	11
0	16.3	100	100	100	100	100	100	100	100	100	57	100
1	73.3	100	100	100	100	100	88	100	100	100	36	100
2	89.7	100	100	100	100	100	78	100	100	100	30	100
3	97.9	100	100	100	88	100	68	100	100	100	26	100
4	99.8	100	100	100	75	100	58	100	100	100	23	50

[2] If the requirements can be met with the largest EV-contributing ATOs removed or failed, the requirements can certainly be met with the lesser EV-contributing ATOs removed or failed.

then hypothesized that the same is true when each of the ATOs has a 90-percent, as opposed to a 100-percent, success rate. We ran the simulation model and found that this hypothesis is valid. The feasible percentage increases from 16.3 percent in row 1 to 73.3 percent in row 2. By the same token, even if the largest four ATO contributors to each FOC were to fail, the remaining ATOs (with a 100-percent success rate) could still meet FOC 4's gap at 75 percent, FOC 6's gap at 58 percent, FOC 10's gap at 23 percent, FOC 11's gap at 50 percent, and all other FOC gaps at 100 percent. When we actually ran our model with a 90-percent success rate for each of the 75 ATOs, there was a 99.8-percent chance of meeting requirements at these levels with the existing ATOs (row 5). With the feasible percentage at essentially 100 percent, we can stop the search for the most cost-effective requirement levels for the existing ATOs to meet.[3] This optimal search also identifies FOCs 4, 6, 10, and 11 as being at risk and, therefore, candidates that the new ATOs should be tailored to meet.

It is of interest to know to what extent planners can reallocate some of the funds for supporting the continuation of existing ATOs to funding new ATOs. An attractive approach is to terminate some existing ATOs so that the overmet FOCs will be much less overmet but the FOC requirements will still be met. The money saved from not continuing the funding of some existing ATOs can then be used to fund new ATOs. This approach will be studied in the section below.

---

[3] We propose a multistep procedure to locate the most cost-effective requirements for the existing ATOs to meet. Since we would want the existing ATOs to meet as many FOC requirements and at as high a feasible percentage as possible, we introduce a metric, the expected total feasible value, which adds all 11 FOC requirement levels and multiplies by the feasible percentage. The metric values for rows 1 to 5 are 172, 751, 904, 961, and 903, respectively. If planners were to base the choice of row solely on the metric value, they would select the requirement levels in row 4 as the most cost-effective requirement levels for the existing ATOs to meet. However, this metric is based on the assumption that the new ATOs needed to fill the remaining requirements unfilled by the existing ATOs would, on average, have the same cost and contribution per ATO as those of an average existing ATO. The assumption could be reasonable but needs to be verified using actual data. To accomplish this, we suggest the following process: Use all rows of similar value as the starting point—in this case, rows 3, 4, and 5. For each row, design new ATOs to actually fill the remaining requirements. Finally, compare the cost-effectiveness of all three rows, each of which will include both new and existing ATOs.

There is an additional consideration. Because it generally takes additional time to design, develop, and complete new ATOs, compared with simply completing existing ATOs, it is important that the requirements planner and the S&T program manager discuss how requirements can be met by existing or new ATOs and whether the fulfillment of a requirement can be delayed until the systems from the new ATOs are ready for fielding. Our model can easily accommodate requirements that cannot wait until new ATOs are developed by insisting that such requirements be met with existing ATOs.

Finally, if planners were forced to choose a row without any knowledge of the designs of new ATOs (the case at hand, since such designs are outside the scope of this study), planners might prefer that the existing ATOs attain a higher feasible percentage rather than fill more FOC requirements. This is because it is easier to design new ATOs to fill FOC requirements than to attain a high feasible percentage. Since the metric values are similar for rows 3–5, we therefore use row 5 as the reference case for this study in order to demonstrate the methodology. Were the designs of new ATOs for filling requirement gaps available, planners would include all three rows of requirements in the comparative analysis.

Determine Total Remaining Lifecycle Budget for Existing ATOs

Based on the lifecycle cost estimates of Appendix B, if all 75 existing ATOs were successfully completed and their systems were all fielded, the total remaining lifecycle cost would be about $138 billion. However, we need not expect the budget to be this high, for two reasons. First, since each ATO is assumed to have a 10-percent failure rate, there is only a 0.04 percent-chance (i.e., $[0.9]^{75}$) that all ATOs would turn out to be successful or that all their systems would be developed and fielded. Second, and much more important, we do not need to field systems from all successful ATOs to meet the 11 FOC requirements because, as shown in Figure 4.2, the successful systems together have the potential (the red bars) to meet requirements far exceeding those required (the dark blue bars). In other words, there are ample redundancies to overcompensate for the inevitable failure of some ATOs if all 75 ATOs are funded to completion. This is confirmed by our simulation, which found that a budget of $67 billion could do practically as well on the likelihood that the reference FOC requirements can be met (feasible percentage) as a budget of $138 billion (no RLCC limit). In fact, the two curves are indistinguishable in Figure 4.3.

That the $67 billion curve and the $138 billion curve are indistinguishable calls for an explanation. Based on the discussion of the simulation in the section "Simulation" in Chapter Two, we can see that the simulation is designed to mimic how the events will unfold and suggest how a decision should be made without knowing what actually will happen in the future. The decision at hand is to select which of the exist-

Figure 4.2
FOC Requirements That Existing ATOs Can Meet with a High Feasible Percentage

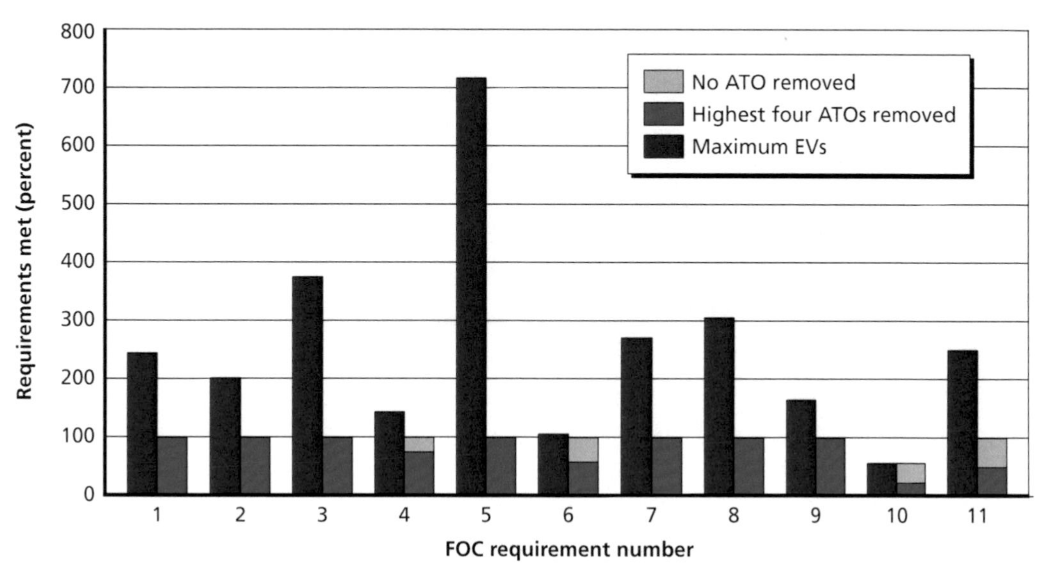

RAND *MG979-4.2*

Figure 4.3
The Chance to Meet All Gaps Within a Total Remaining Lifecycle Cost Limit

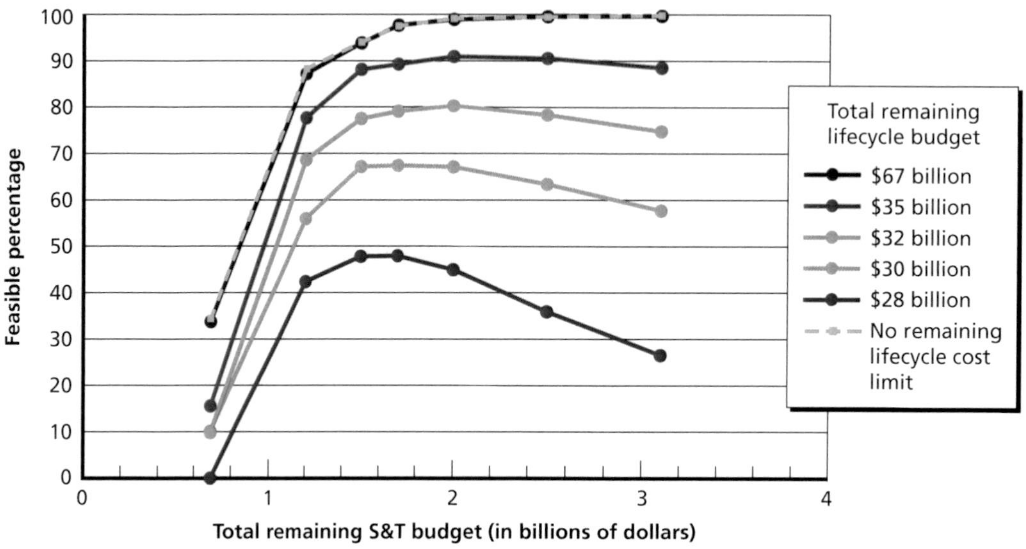

RAND *MG979-4.3*

ing 75 ATOs should be chosen for continued funding until their completion. The aim is to select the optimal portfolio that has the highest chance (feasible percentage) to meet all requirements within the given budgets for total remaining S&T and total remaining lifecycle costs.[4] We have developed a search algorithm to find the optimal portfolio, which is the portfolio with the highest feasible percentage among all the trial portfolios. A trial portfolio is a specific selection of the existing 75 ATOs. For every trial portfolio, we run the simulation 10,000 times; each run is a set of independent random draws: 90-percent chance of success and 10-percent chance of failure for every selected ATO.[5] For each run, there will be a set of successfully completed ATOs resulting from the random draws. The simulation will determine whether these successful ATOs can meet all requirements within budget constraints. If so, it is a *feasible* run. The simulation will then tally all the feasible runs and express them as a percentage of the 10,000 runs. This percentage is the feasible percentage for the trial portfolio. Our search algorithm aims to find a trial run with a higher feasible percentage than the previous trial run, thus efficiently and quickly arriving at the optimal portfolio. That the feasible percentage is practically the same[6] whether the total remaining lifecycle cost is

---

[4] Recall that the total remaining lifecycle cost is the sum of the total remaining S&T cost and the total implementation cost. The latter includes the cost for engineering and manufacturing development, procurement of multiple copies of systems, and the operating and maintenance of these fielded systems.

[5] The model allows for different ATOs to have different probabilities of success.

[6] The feasible percentage for the $138 billion is actually a fraction of a percent higher.

$67 billion or $138 billion means that the binding constraint is in the total remaining S&T budget. Sixty-seven billion dollars is enough to pay for total remaining S&T cost and the total implementation cost in almost every case that the successful ATOs are sufficient to meet all requirements. From a different perspective, we can also say that the ATOs not selected for the optimal portfolio are those of low EV contribution but high S&T cost and high implementation cost. These ATOs are not picked because they are inferior to or dominated by the ones in the optimal portfolio, where a total remaining lifecycle cost of $67 billion is almost always all that is needed. The extra $71 billion (i.e., $138 billion minus $67 billion) is not needed in practically all the cases.

The next question is whether we should budget $67 billion as the total remaining lifecycle cost for the existing 75 ATOs. This funding level allows close to 100-percent chance or certainty (actually 99.8 percent)[7] that all the requirements will be met. Figure 4.3 also shows that a $35 billion budget would yield a 90.9-percent probability of meeting all requirements.[8] Because the $67-billion budget would yield a nearly 100-percent chance of meeting all requirements, a linear approximation would mean that it takes, on average, $0.67 billion to raise the chance by one percentage point. This offers the option of budgeting only $35 billion for the existing ATOs but using part of the $32 billion in savings (i.e., $67 billion minus $35 billion) to fund new ATOs and their systems so as to raise the chance from 90.9 to 99.8 percent. Raising the chance by 8.9 percentage points would require only about $6.0 billion (i.e., $0.67 × 8.9), which is only a small portion of the $32 billion, for a net savings of $26.0 billion or 81 percent (i.e., 26.0 ÷ 32).[9] While a linear approximation is not precise, it clearly indicates that the program manager of the ATO portfolio should study the designs and costs of new ATOs and compare the two options for meeting the requirements at a high feasible percentage:

1. Continue funding most of the existing ATOs to their completion and deploy some or all of the successful ATOs' systems to meet requirements.
2. Reduce funding to the existing ATOs and their systems and use the savings to support new ATOs and their systems.

---

[7] In this monograph, we often show numbers beyond their number of significant figures. The purpose is to make it easier for interested readers to follow and reproduce our calculations.

[8] We assume that the total remaining S&T budget is set at $2.0 billion. Moreover, the results are similar if the budget is somewhere between $1.5 billion and $3.1 billion—the curve is flat in this region, as shown in Figure 4.3.

[9] In order for the $6.0-billion budget to be adequate, we must be able to identify which requirements are most difficult for the existing ATOs to meet so that the new ATOs can be tailored to meet these requirements. Planners can use our model to examine those 8.9-percent cases in which all the requirements are not met and to identify which requirements are not met in those 8.9-percent cases. Only then can planners design new ATOs specifically to meet these identified requirements cheaply, as discussed in the section "Establish Realistic FOC Requirements for Existing ATOs."

Our simulation can quantify the trade-off and choose the less expensive option. As a rule of thumb, planners should consider reducing the feasible percentage attained by the existing ATOs in order to produce savings in the TRLCC for these existing ATOs, provided that the TRLCC for the new ATOs to recover the lost feasible percentage and to attain the old feasible percentage is less than the savings.[10] In the above case, reducing the feasible percentage from 99.8 to 90.9 percent would save $32 billion, and yet adding new ATOs so as to reach 99.8 percent again would cost only $6.0 billion, which would be much less than $32 billion. In such a case, we would recommend reducing the total remaining lifecycle budget for the existing ATOs from $67 billion to $35 billion and using $6.0 billion of the $32 billion savings to fund new ATOs and their systems in order to return to a feasible percentage of 99.8, resulting in a net savings of $26.0 billion. This rule of thumb should be useful when deciding on the distribution of funds between existing and new ATOs.

On the other hand, reducing the TRLCC further from $35 billion to $32 billion would save only $3 billion. Yet, the chance of meeting all gaps would drop from 90.9 to 80.3 percent, or a drop of 10.6 percent. Funding new ATOs and their systems to return to 90.9 percent would require $7.1 billion (i.e., $0.67 × 10.6). Since $7.1 billion is higher than the initial savings of $3 billion, the $35 billion TRLCC budget for the existing ATOs should not be reduced in order to fund new ATOs.

Similarly, further reducing the TRLCC from $32 billion to $30 billion for a $2 billion savings would lead to a drop in feasible percentage from 80.3 to 67.4 percent, or a 12.9-percent drop. Recovering the 12.9-percent drop would cost $8.7 billion (i.e., $0.67 × 12.9), which is certainly not a cost-effective trade.

Finally, reducing from $30 billion to $28 billion would result in a $2-billion initial savings but would lead to a drop in feasible percentage from 67.4 to 47.9 percent or a 19.5-percent drop. It would cost $13.1 billion (i.e., 0.67 × 19.5), which is much larger than the $2-billion initial savings, and thus, going from $30 billion to $28 billion is even more cost-ineffective.

For this study, we chose the reference budget for total remaining lifecycle cost for the existing 75 ATOs to be $35 billion, shown as the red line in Figure 4.3.

Determine the Optimal S&T Budget

The largest feasible percentage attainable depends on the amount of funds available for completing the ATOs and implementing their systems. Figure 4.3 shows the largest feasible percentages for various constraints in the TRSTC and TRLCC. As the TRLCC consists of the TRSTC and the total marginal implementation cost (TMIC),

[10] As discussed in the previous section, we should also consider that certain requirements must be met with existing ATOs if new ATOs would take longer to develop systems for meeting these requirements.

the analysis below shows how a given TRLCC budget should be allocated between TRSTC and TMIC to yield the highest chance (feasible percentage) of meeting all requirements.

When there is no constraint on TRLCC (i.e., $138 billion, the cost of funding all 75 ATOs and implementing all their systems), the feasible percentage is always largest when there is no constraint on TRSTC. Stated differently, a limited TRSTC can possibly yield a larger feasible percentage, when the TMIC budget is constrained to be unable to support the fielding of all the systems. Then, when planners limit TRSTC spending and transfer the saved TRSTC funds to the TMIC budget, the feasible percentage may increase through relaxing the TMIC constraint.

As shown in the no-TRLCC-limit cases, the feasible percentage decreases monotonically when the TRSTC becomes more and more limited. In other words, since TRLCC is unlimited and there are unlimited funds for continuing all ATOs and fielding their systems, setting a limit on TRSTC can only hurt the feasible percentage.

For the reference TRLCC of $35 billion, the model suggests that the optimal TRSTC is near $2.0 billion (see Figure 4.3). In contrast, the TRSTC needed to fund all existing 75 ATOs to their completion is $3.1 billion. The simulation says that if planners spend $2.0 billion of the $35 billion TRLCC on TRSTC, they can expect a 90.9-percent chance of meeting all requirements. If planners spend $3.1 billion on TRSTC and $1.1 billion (i.e., 3.1 – 2.0) less on system implementation, they can expect an 88.5-percent feasible percentage. Thus, the allocation of $2.0 billion of the $35 billion TRLCC to TRSTC is preferred, because it yields a higher feasible percentage for the same $35 billion spending.[11] For this study, we use the $2.0 billion TRSTC and $35 billion TRLCC as the reference case (see Figure 4.3). Moreover, we consider the reference case to be at the sweet spot, because it is located at the most cost-effective combination of the total S&T budget (TRSTC) and the remaining lifecycle budget (TRLCC).

Select ATOs for Continued Funding

There are reasons that some of the existing ATOs may be discontinued. For example, an economic crisis might force budgetary cuts across the board, including funds supporting the existing ATOs. Also, even when the S&T budget is unchanged, the FOC requirements might change, so that some existing ATOs no longer serve their purposes. It might become clear that certain existing ATOs can no longer meet their project objectives in developing systems of adequate performance and affordable cost. Then, it might be better to replace these poorly performing ATOs with new ones. The

[11] We discuss the ramifications of the flat region between $1.5 billion and $3.1 billion in the section "Impact of Suboptimal Total Remaining S&T Budget."

Table 4.2
Optimal Subset of ATOs That Produces the Largest Feasible Percentage When the Total Remaining Lifecycle Budget Is $35 Billion

S&T \ ATO	1	2	3	4	5	6	7	8	9	10	11	12	13	14	15	16	17	18	19	20	21	22	23	24	25	26	27	28	29	30	31	32	33	34	35	36	37	38
$ 3.1 B	Y	Y	Y	Y	Y	Y	Y	Y	Y	Y	Y	Y	Y	Y	Y	Y	Y	Y	Y	Y	Y	Y	Y	Y	Y	Y	Y	Y	Y	Y	Y	Y	Y	Y	Y	Y	Y	Y
$ 2.5 B	Y	Y	Y	Y	Y	Y	Y	N	Y	Y	Y	N	Y	Y	N	N	Y	Y	Y	N	Y	Y	Y	Y	N	Y	Y	Y	Y	Y	Y	Y	Y	Y	N	Y	Y	Y
$ 2.0 B	Y	Y	Y	Y	N	Y	Y	N	Y	Y	Y	N	N	Y	N	N	N	Y	Y	N	N	Y	Y	Y	N	Y	Y	Y	Y	Y	Y	N	Y	Y	N	Y	Y	Y
$ 1.7 B	Y	Y	Y	Y	N	Y	Y	N	Y	N	N	N	Y	Y	N	N	N	Y	Y	N	N	Y	Y	Y	N	Y	Y	Y	Y	Y	Y	N	Y	Y	N	Y	Y	Y
$ 1.5 B	Y	Y	Y	N	N	Y	Y	N	Y	Y	N	N	Y	Y	N	N	N	Y	Y	N	N	Y	Y	Y	N	Y	Y	Y	Y	Y	Y	N	Y	Y	N	Y	Y	Y
$ 1.2 B	Y	Y	Y	N	N	Y	Y	N	Y	N	N	N	Y	Y	N	N	N	Y	Y	N	N	Y	Y	Y	N	Y	Y	Y	Y	N	Y	N	N	Y	N	Y	N	Y

S&T \ ATO	39	40	41	42	43	44	45	46	47	48	49	50	51	52	53	54	55	56	57	58	59	60	61	62	63	64	65	66	67	68	69	70	71	72	73	74	75	#
$ 3.1 B	Y	Y	Y	Y	Y	Y	Y	Y	Y	Y	Y	Y	Y	Y	Y	Y	Y	Y	Y	Y	Y	Y	Y	Y	Y	Y	Y	Y	Y	Y	Y	Y	Y	Y	Y	Y	Y	75
$ 2.5 B	Y	Y	Y	Y	Y	Y	Y	Y	Y	Y	Y	Y	Y	Y	Y	Y	Y	Y	Y	N	Y	Y	Y	Y	N	Y	Y	N	Y	Y	Y	Y	N	Y	Y	Y	Y	64
$ 2.0 B	N	Y	Y	Y	Y	Y	N	Y	Y	Y	Y	Y	N	Y	Y	Y	N	Y	N	N	Y	Y	Y	N	N	Y	Y	N	Y	Y	Y	Y	N	Y	Y	Y	Y	53
$ 1.7 B	N	Y	Y	Y	Y	Y	N	Y	Y	N	Y	N	N	Y	Y	Y	Y	Y	N	N	Y	Y	Y	Y	N	Y	Y	N	Y	Y	Y	Y	N	Y	Y	Y	Y	52
$ 1.5 B	N	N	Y	Y	Y	Y	N	N	Y	N	Y	N	N	Y	Y	Y	Y	Y	N	N	Y	Y	Y	N	N	Y	Y	Y	Y	Y	Y	Y	N	Y	Y	Y	Y	50
$ 1.2 B	N	N	Y	N	Y	Y	N	N	N	N	Y	N	N	Y	Y	Y	N	Y	N	N	Y	Y	Y	N	N	Y	Y	N	Y	Y	Y	Y	N	Y	Y	Y	Y	42

NOTE: *Y* signifies that the ATO will continue to be funded, and *N* that it will be terminated. The last column shows the total number of ATOs funded.

model developed in this study can analyze these situations and recommend a cost-effective course of action.

For illustration, we used the model to address the following question:

> Given that the total remaining lifecycle cost is restricted to not exceeding $35 billion, which of the 75 existing ATOs should be kept in order to have the highest feasible percentage, or chance, of meeting the 11 FOC requirements?

We first followed the line of thinking in the previous section to determine the optimal allocation of funds between total remaining S&T budget and total marginal implementation budget. The previous section found that the allocation yielding the highest chance is $2.0 billion for TRSTC, as shown in Figure 4.3.

The simulation searches for a subset of the 75 ATOs so that the feasible percentage is maximized. The row highlighted in yellow in Table 4.2 shows the optimal set for continued funding to consist of 53 ATOs. Twenty-two ATOs (indicated by an *N* on the highlighted row) are not selected because we can attain a higher feasible percentage by using the money saved from discontinuing them to support the implementation of systems from the ATOs that continue to be funded and turn out to be successful.

There may be economically stressful situations in which DoD could not support the ATOs at the optimal level ($2.0 billion). Our model can determine the best subset of ATOs to be kept and which ATOs to be terminated under various budget levels. Both classes of ATOs are also shown in Table 4.2 for each budget level. The Deputy Secretary of Defense issued directives in 2006 and 2008 to use capability portfolio management for planning and implementing capability development. The tools developed here can be useful for Army and other services to perform CPMT, particularly in identifying which ongoing projects to trim when they face a budget cut, which is a frequent occurrence in the current economic climate. Decisionmakers might be curious whether the model provides keep and/or terminate decisions for the 75 existing ATOs that other much simpler rules can replicate. We have devised several such simpler rules. The best of these is based on an ATO's total expected value of contributions to all FOCs, divided by the remaining S&T cost of that ATO (RSTC). A high ratio for this indicator means that an ATO contributes a lot to requirements, but costs only a little. Thus, an ATO with a high TEV-RSTC ratio is attractive, and we would expect that it should be kept for further funding. Figure 4.4 shows that, while ATOs with higher ratios (green bars) are mostly selected for continued funding, and those with lower ratios mostly terminated, there are several important exceptions in the optimal portfolio. For example, the TEV-RSTC ratio cannot predict that ATOs such as 58, 13, and 71 (in red) should be rejected although they are interspersed among the selected ones (in green). Our model rejects these ATOs because they contribute to FOC gaps that are already met more efficiently by other ATOs. Moreover, the TEV-RSTC ratio cannot predict that some ATOs—such as 48, 46, and 47—with lower ratios than several rejected ATOs should still be kept. Although these ATOs do not have attractive

Figure 4.4
ATO Ordering According to the Ratio of Total Expected Value over the Remaining S&T Cost

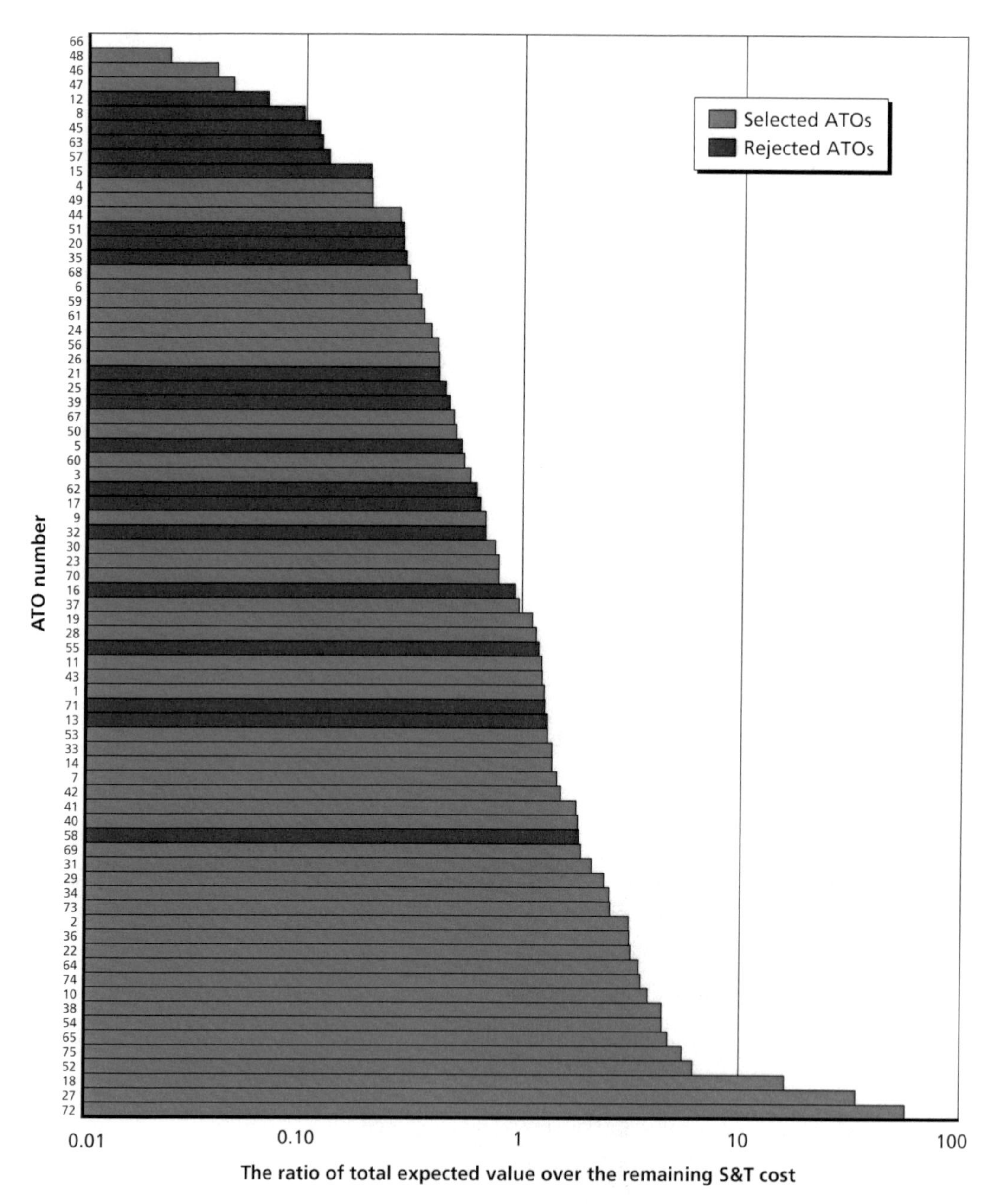

RAND *MG979-4.4*

TEV-RSTC ratios, they should be included in the optimal portfolio in order to fully meet certain FOC gaps. Our model automatically takes these considerations into account.

Figure 4.5 shows our three other attempts to develop simple indicators for the selection of ATOs for continued funding. The top panel is a plot of an ATO's total EV against its remaining S&T cost. The mixing of the selected ATOs (green) and the rejected ones (red) or the lack of isolated regions for greens only or reds only (such as the red rectangles in Figure 4.5) prevents the plot from being a useful rule. Similarly, a plot of an ATO's total EV against its marginal implementation cost and a plot of an ATO's remaining S&T cost against its marginal implementation cost do not yield a useful rule for selecting ATOs for continued funding. Thus, we conclude that it is difficult, and perhaps impossible, to devise any simple rule to help decisionmakers select which ATOs should be discontinued. A model is needed to deal with the complicated interactions among ATOs and requirements and to determine which ATOs should be discontinued.

Impact of Suboptimal Total Remaining S&T Budget

While the Army should try to keep the remaining S&T budget for the existing ATOs at the optimal level, as discussed in the section "Determine the Optimal S&T Budget" above, it, like other services and departments, faces unexpected budgetary cuts from time to time. It is important to know the impacts of such cuts so that ATO program managers can communicate with their superiors and eventually arrive at cuts that provide the best compromise that meets essential capability requirements consistent with the current budgetary reality. Our model quantifies the impacts and suggests a cost-effective compromise.

There are two types of cuts in the total remaining S&T budget for the existing ATOs. A type-1 cut allows the lost funding to be recovered and used later for implementation (i.e., demonstrating, acquiring, fielding, and operating systems). A type-2 cut in the remaining S&T budget is never recovered. We address each type of cut separately in the following subsections.

A Type-1 Cut in the Total Remaining S&T Budget

Figure 4.3 can be used to determine the impact from cuts of the first type. For example, in the case of $35 billion TRLCC, a cut of TRSTC to $1.7 billion would reduce the feasible percentage to 89.3 percent (from 90.9 percent at $2.0 billion); a cut to $1.5 billion, 88.1 percent; and a cut to $1.2 billion, 77.6 percent. If a budget cut in TRSTC brings the feasible percentage to the sharply declining region, as in the last case above, the ATO program managers should alert their superiors to the consequent

Figure 4.5
Simple Indicators Without the Model Are Inadequate for Selection Decisions

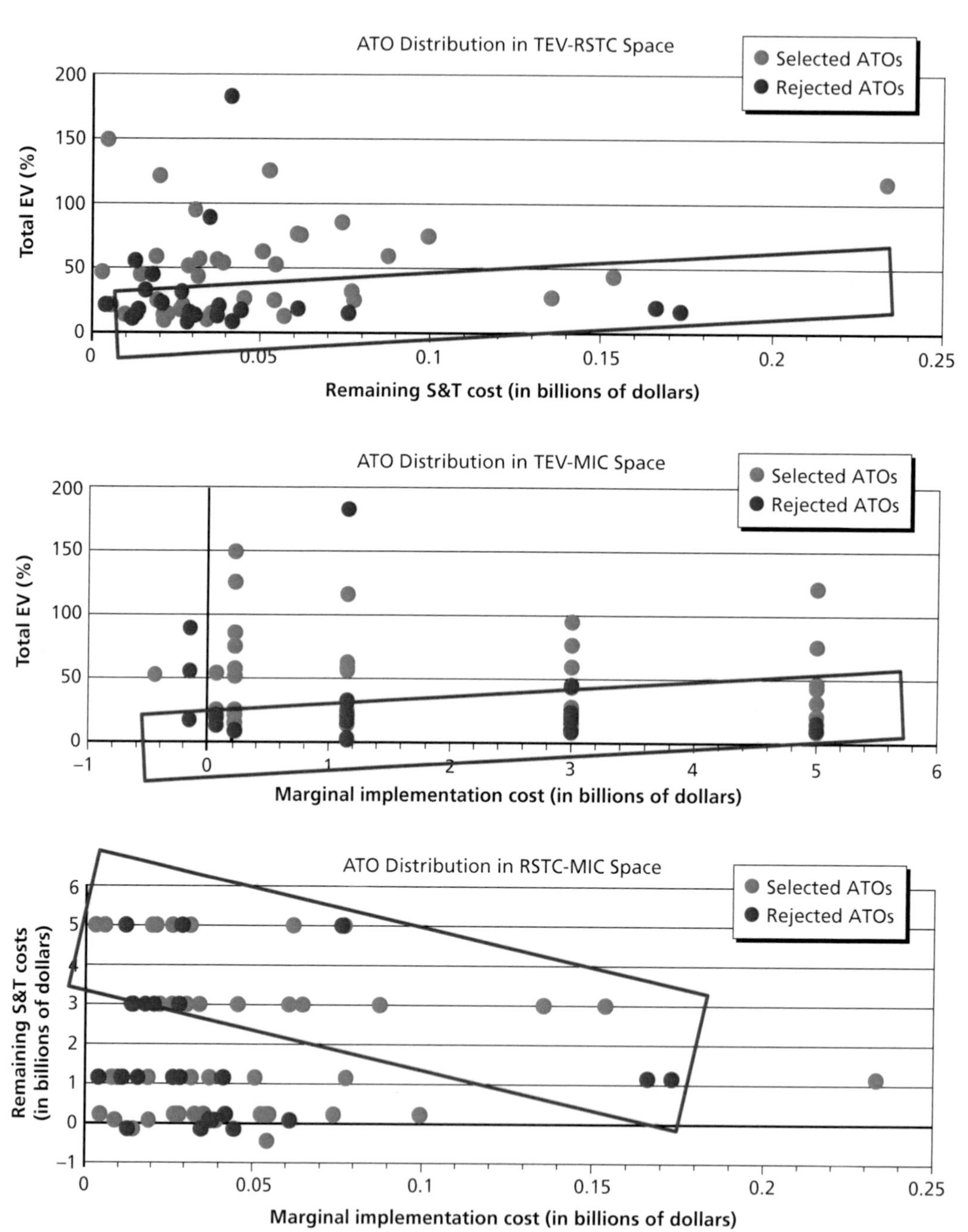

RAND *MG979-4.5*

drastic decline in the chance of meeting future FOC requirements. Knowing the severe impacts, the superiors may elect to reduce the cut to a much less damaging level.

As shown in Figure 4.3, the sharp drop in feasible percentage occurs at around $1.5 billion in total remaining S&T budget, regardless of total remaining lifecycle budget. Moreover, the feasible percentage is relatively insensitive to the S&T budget between $1.5 billion and $3.1 billion, as long as the total remaining lifecycle cost budget remains at $35 billion. In other words, as long as the saved S&T money is recovered and used for implementation, it is a type-1 cut.

A Type-2 Cut in the Total Remaining S&T Budget

We have made simulation runs additional to those shown in Figure 4.3 in order to address cuts of the second type. For example, if the TRSTC is cut from $2.0 billion to $1.5 billion and the money is never recovered, the total remaining lifecycle budget (because it includes the TRSTC) would be similarly cut—i.e., from $35 billion to $34.5 billion. The feasible percentage would be reduced to 87.2 percent, which is less than the 88.1 percent when the $0.5 billion is recovered. Thus, a type-2 cut will always reduce the feasible percentage more than a type-1 cut. In other words, if the degradation in feasible percentage is serious under a type-1 cut, a type-2 cut of the same amount in total remaining S&T budget will be even worse. Again, planners need to ascertain the type of cut and to determine its impact, so that the final decision on the cut is made with the full knowledge of the cut's consequences.

It should also be noted that an S&T budget cut that is not recovered will always reduce the feasible percentage. However, this is not explicitly shown in Figure 4.3, which only shows recoverable S&T budget cuts. On the other hand, the feasible percentage reduction can easily be calculated by interpolation using neighboring total remaining lifecycle curves, as demonstrated in the discussion above.

Optimal Distribution of Funds Among FOC Gaps

Figure 2.2 shows the remaining S&T costs of the 75 ATOs, while Figure 2.3 shows their marginal implementation costs. Further, Table 4.2 shows that there are 53 ATOs that should continue to be funded in order to have the highest chance to meet all requirements (dark blue bars in Figure 4.2) within a total remaining lifecycle budget of $35 billion ($2 billion for total remaining S&T cost and $33 billion for TMIC). This section addresses the remaining S&T cost and marginal implementation cost required to meet each of the 113 individual gaps from TRADOC/ARCIC.

Allocating cost among gaps is complicated by two commonly occurring situations:

- A single ATO can contribute to multiple gaps.
- Multiple ATOs can contribute to the same gap.

We first use a simplified example to demonstrate our scheme of cost allocation, which is based on proportions. As shown in Table 4.3, the simplified example consists of only two ongoing ATOs, which are to develop systems to meet two gaps. We start by assuming that the requirements are to meet gap 1 at 100 percent and gap 2 at 75 percent. As it turns out, our scheme is independent of the levels of requirements to meet, so long as it is equally important to meet 100 percent of gap 1 and 75 percent of gap 2.[12] Further, we assume that ATO 1's EV for gap 1 is 75 percent, and its EV for gap 2 is 50 percent. ATO 2's EV for gap 1 is assumed to be 40 percent and its EV for gap 2, 80 percent. We

Table 4.3
Simplified Example to Demonstrate This Study's Cost Allocation Scheme

Individual Gap Requirements

	Gap1	Gap2
Requirement	100%	75%

ATO Absolute Contributions

	Gap1	Gap2
ATO1	75%	50%
ATO2	40%	80%

ATO Relative Contributions

	Gap1	Gap2
ATO1	65%	38%
ATO1	35%	62%

ATO Cost Allocations

	Gap1	Gap2
ATO1 ($3M)	$1.9M	$1.1M
ATO2 ($5M)	$1.8M	$3.2M
	$3.7M	$4.3M

[12] For example, planners recognize that gap 2 is much more expensive to meet and feel that meeting gap 2 at 75 percent is just as important as meeting gap 1 at 100 percent. On the other hand, our scheme can be easily modified to treat meeting gap 2 at 100 percent as equally important as meeting gap 1 at 100 percent. For example, we can use a weight factor of 75 percent when gap 2 is met at 75 percent.

also assume that the remaining S&T cost for ATO 1 is $3 million and that for ATO 2 is $5 million. The allocation procedure then executes the following steps:

- Determine the relative contribution of ATO 1 to gap 1 as 65 percent (i.e., 75 percent ÷ [75 percent + 40 percent]). This means that we weigh the ATO 1 contribution relative to those of other ATOs to the same gap. Using the same approach, we see that the relative contribution of ATO 1 to gap 2 is 38 percent (i.e., 50 percent ÷ [50 percent + 80 percent]).
- Allocate the remaining S&T cost to fill gap 1 from ATO 1 as $1.9 million—i.e., $3 million × 65 percent ÷ (65 percent + 38 percent). This means that we allocate the costs of ATO 1 to the two gaps according to its relative contributions to them. Using the same approach, we see that the remaining S&T cost to fill gap 1 from ATO 2 is $1.8 million.
- Determine the cost to fill gap 1 as $3.7 million (i.e., $1.9 million + $1.8 million). This is simply a sum of the allocated costs from ATO 1 and ATO 2.

Applying this proportional scheme to the 53 continuing ATOs (but excluding the 22 terminated ATOs because no more S&T funds would be spent on them) and the 113 gaps, we obtained the allocation of total remaining S&T cost of $2 billion among these gaps (Figure 4.6). It is clear that funding all gaps equally is suboptimal, since the

Figure 4.6
Remaining S&T Costs to Meet Individual Capability Gaps

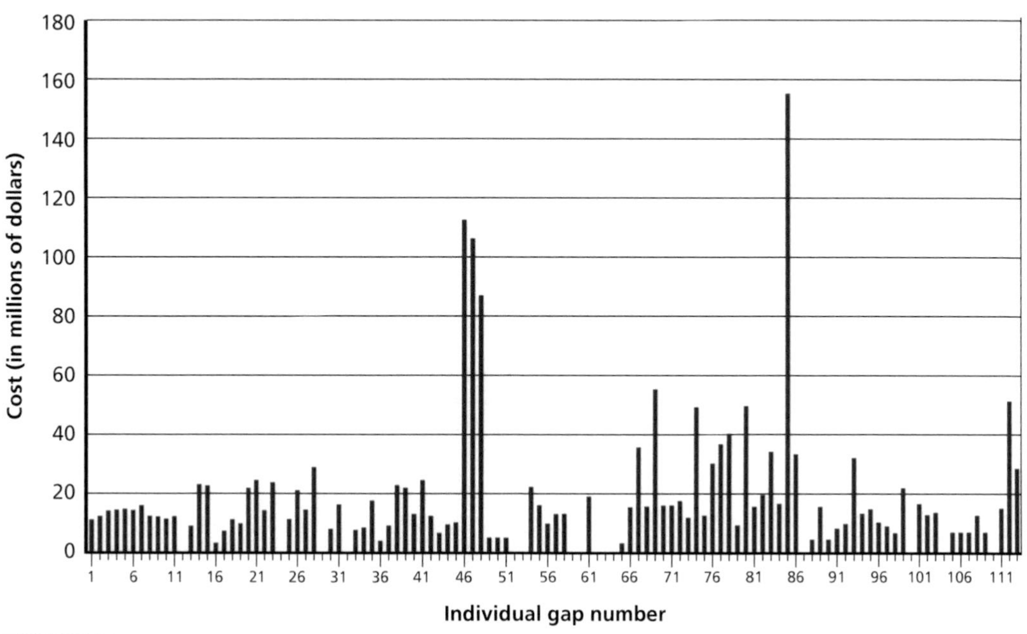

cost to develop the systems for meeting a gap varies among gaps. Moreover, because of the two situations described at the beginning of this section, we need to cost all gaps at the same time.

Allocating the expected TMIC of $22 billion[13] among the gaps is more complicated than the remaining S&T cost allocation. Planners need to first review the decision process. For the reference case, 53 ATOs have been selected for continuation. Upon their completion, most of the ATOs will be successful, because the failure rate is only 10 percent. On the other hand, several of them will fail, and the failed ones are determined by random and independent success and/or failure draws of the 53 ATOs individually. For Figure 4.7, we asked the simulation to perform 10,000 runs or sets of draws. For each run, the simulation tracks whether systems from the successful ATOs can meet all 11 FOC requirements. If they do, the simulation further tracks systems from which successful ATOs will be implemented to meet the requirements at the lowest TMIC. The expected marginal cost of implementing the systems from a specific ATO is the product of two factors. Factor 1 is the frequency, or the number of times,

Figure 4.7
Expected Marginal Implementation Costs to Meet Individual Capability Gaps

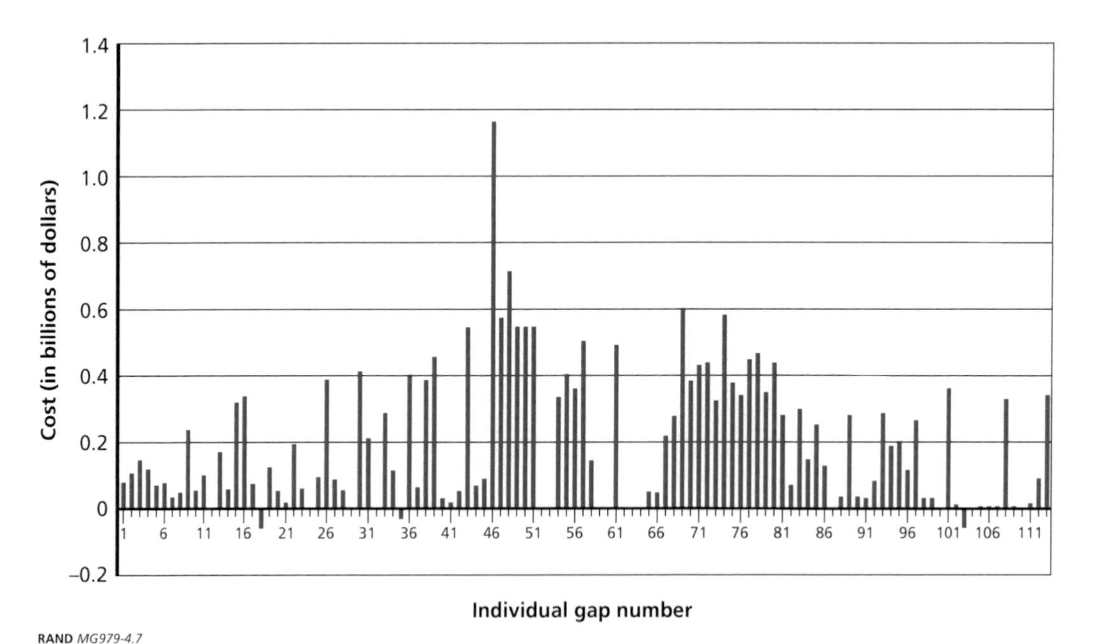

RAND *MG979-4.7*

[13] With a total marginal implementation budget of $33 billion, the optimal portfolio can meet all 11 FOC requirements in 91 percent of the possible future cases. Fortunately, in many of those cases, these requirements can be met at budgets much lower than $33 billion. In other words, in only the most stressful cases, where many of the ATOs fail, will the backup ATOs with much more expensive implementation costs be needed, and the total remaining implementation cost would be close to $33 billion. The expected or average total remaining implementation cost turns out to be $22 billion.

that a given ATO's systems participate in runs for which the 11 FOC requirements are met divided by the number of runs for which the requirements are met (with and without the participation of the ATO in question). Factor 2 is the cost of implementing systems derived from the ATO in question. Finally, the expected marginal implementation cost for each gap is determined by the same scheme as that of the remaining S&T cost described above, except that the remaining S&T cost is replaced by the expected marginal implementation cost for each of the 53 ATOs. Figure 4.7 shows the expected marginal implementation cost to fill each gap when systems from the successful ATOs are able to meet all 11 FOC requirements.

To allocate the cost to the 11 FOC requirements, we simply add all the costs of the individual gaps belonging to their respective FOCs. The allocation of remaining S&T cost to meet gaps in an FOC is shown in Figure 4.8. On the lower panel, the cost for each FOC is divided by the number of gaps in that FOC, reflecting the cost per gap in each FOC. A gap within FOC 4 (air maneuver) and FOC 5 (LOS/BLOS/NLOS lethality) requires a much higher S&T budget to fill. Again, this indicates that a uniform allocation of S&T funds for the 11 FOCs is suboptimal, as some FOC gaps can be inherently more expensive to meet. It is also suboptimal to allocate funds based on how many gaps are in each FOC.

Figure 4.9 shows the expected marginal implementation cost to meet gaps in each FOC. Again, it is important to consider the inherent expensiveness of ATO systems in meeting different FOCs.

We close this chapter with two points about costing a gap. First, because an ATO can contribute to multiple gaps and multiple ATOs can contribute to the same gap, we need to cost all gaps at the same time. Second, although we can allocate costs reasonably among gaps and determine the cost to meet each gap, these individual costs are nowhere near as useful as grouped costs. For example, if planners want to decide whether they can afford to meet gap A, gap B, or both, the best approach is to team gap A with the rest of the gaps and use our model to determine the cost to meet all of these gaps. Repeat the process with gap B and the rest of the gaps. Finally, repeat the process with gaps A and B and the rest of the gaps. Then and only then do planners know the costs for meeting three requirement cases (gap A, gap B, and gaps A and B), and then they can make an intelligent choice by comparing these costs and their contributions to requirements. Our model can compare the costs and merits of multiple sets of multiple gaps, without the need to allocate the costs to individual gaps in the process, and come to a cost-effective solution. On the other hand, if planners know the costs of these individual gaps but they do not have the model, the planners cannot arrive at such a solution.

Figure 4.8
Remaining S&T Cost to Meet Gaps in Each FOC

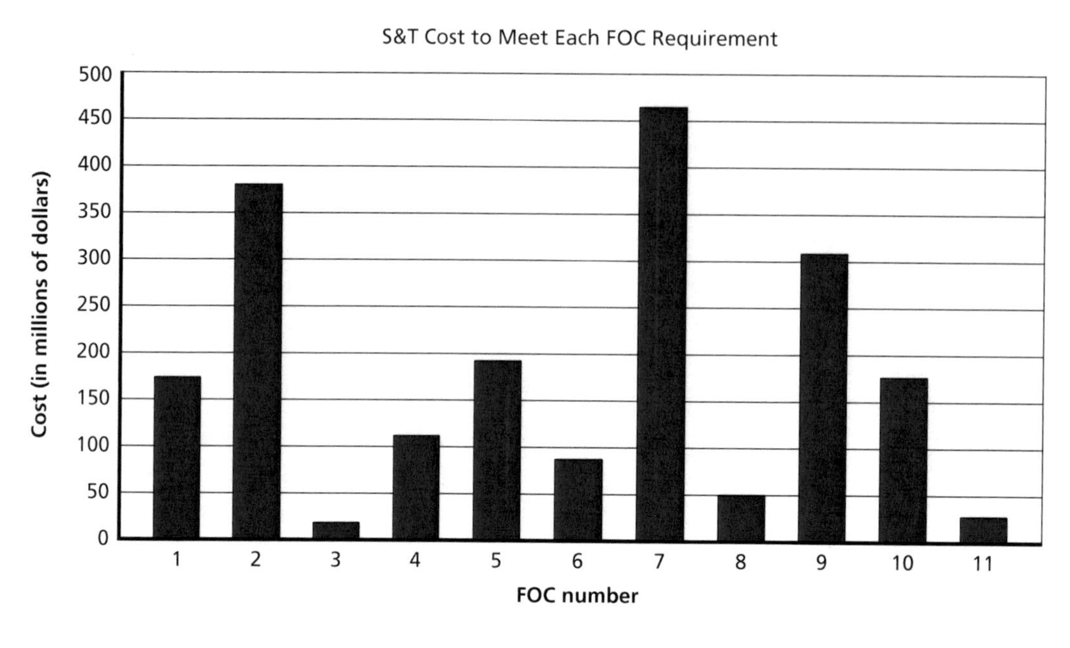

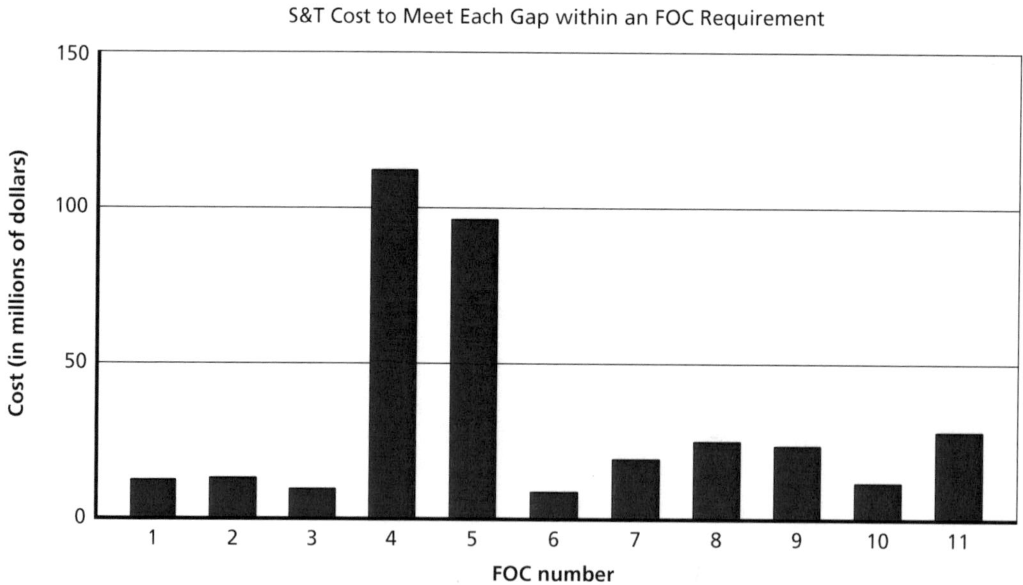

RAND *MG979-4.8*

Figure 4.9
Expected Marginal Implementation Cost to Meet Gaps in Each FOC

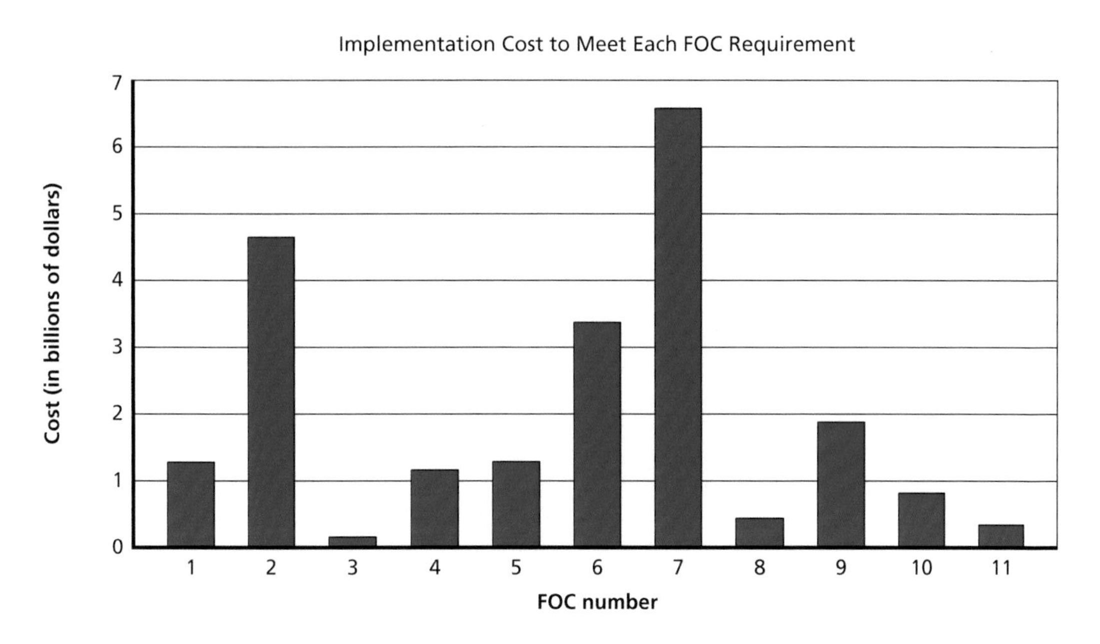

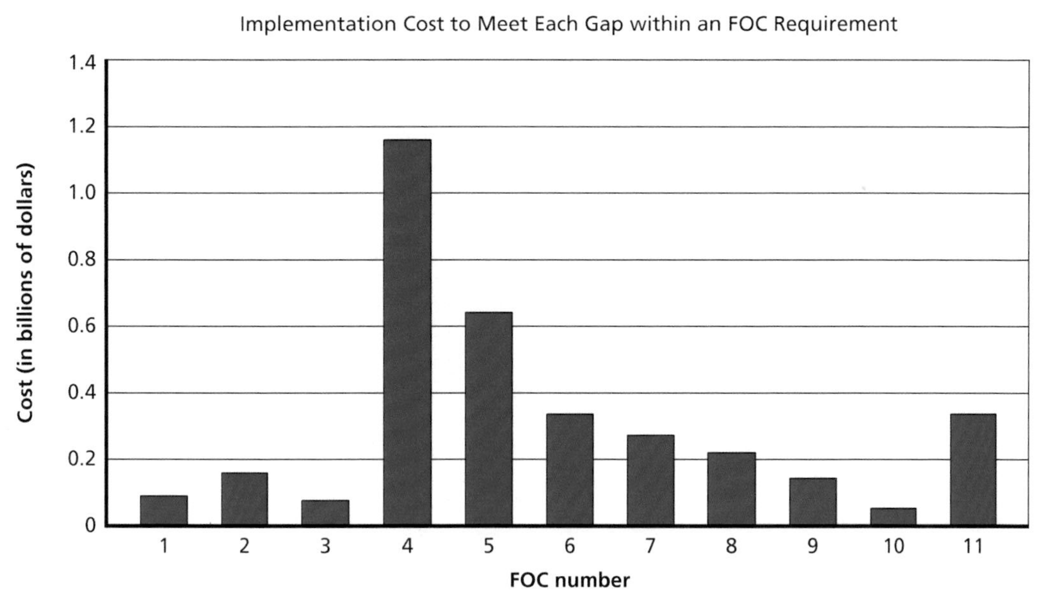

RAND *MG979-4.9*

CHAPTER FIVE

Findings and Recommendations

This chapter describes our findings based on the development and application of the process—which includes the method, model, and simulation shown in the previous chapters—and provides recommendations for use of the process by the Army and other services. Our process can be applied to track and analyze the complicated interactions among driving factors to successfully perform capability portfolio management as required by the Vice Chief of Staff of the Army and DoD.

Findings

We describe eight principal findings below. The first five concern the ability of ATO portfolios to meet FOC capability gap requirements and associated budget issues. The final three are attributes of the model developed and demonstrated in this study. For each one, we refer to the relevant part of the monograph upon which it is based and in which the appropriate application of our process is described.

1. *ATOs are not of equal importance.* As shown in Chapter Three, some ATOs are of special importance because they provide coverage for FOC capability gaps that otherwise have little or no coverage, or provide a substantial fraction of the EV for one or more FOCs. Other ATOs may be less important because they provide a minimal fraction of the EV of any FOC. Our process, as discussed in Chapter Four, which accounts for the complicated interrelations of ATOs in meeting multiple FOCs, can be used to determine the importance of individual ATOs and to upgrade the less important ones or even replace them with better-contributing new ATOs.
2. *Allowing for uncertainty in the success of ATOs makes an important difference in meeting capability requirements.* As shown in Chapter Four, just a 10-percent failure rate across ATOs (90-percent probability of success) results in a mere

16.3 percent probability of meeting all FOC requirements,[1] even with the FOC 10 requirement already reduced to 57 percent (the maximum that could be met with all ATOs at a 100-percent success rate). Our process can be used to determine the impact of the unavoidable failure of some ATOs, which planners cannot predict in advance, and can provide a means to deal with this reality.

3. *A portfolio approach can identify those FOCs that are at risk and those FOCs that are overmet.* As shown in Chapter Four, these insights can allow a more cost-effective allocation of ATO resources to cover all the FOC capability gap requirements. We demonstrated there that defining new ATOs tailored to meet the at-risk requirements would take many fewer new ATOs and cost much less than adding ATOs similar to the existing ones to meet these requirements.
4. *Effects of budget cuts are not linear.* As shown in Chapter Four (see Figure 4.3), changes in the budget for total remaining lifecycle cost have different effects on the probability of meeting all FOC requirements in different cost regions, with much higher loss of cost-effectiveness (as measured by the feasible percentage) in sharply declining regions. This indicates the importance of using a model to reveal the larger picture so that planners know the consequences of a contemplated budget cut.
5. *There is an optimal S&T budget.* As shown in Chapter Four (see Figure 4.3), there is an S&T budget for each remaining total lifecycle cost budget that maximizes the probability that all FOC requirements will be met. There is also a range of S&T budgets in which the change in this probability, as measured by feasible percentage, is small. Thus it is important to determine the change in feasible percentage for any proposed change in S&T budget. Our model can provide these data.
6. *Our model can demonstrate the trade-offs associated with budget cuts and identify the most cost-effective options.* As discussed in Chapter Four, trade-offs between funding new or existing ATOs and those associated with shifting funds between remaining S&T cost and marginal implementation cost, when budgets are cut from the optimum case, are complex and depend on the interplay between the level of FOC requirements to be met and the EV and cost of the ATOs under consideration. Our model combines these factors in a consistent manner, elucidates the trade-offs, and identifies the most cost-effective options at any budget level.
7. *Our model also can indicate which ATOs to keep when the S&T budget chosen is optimal or suboptimal.* As shown in Chapter Four, our model can select the highest feasible percentage portfolio of ATOs for any S&T budget and remain-

[1] Throughout this chapter we use feasible percentage, as defined in the subsection "Simulation" in Chapter Two, as a measure of the probability of meeting all FOC requirements.

ing lifecycle budget, whether they are optimal or suboptimal (see Table 4.2 for a given remaining lifecycle budget of $35 billion as an example).

8. *Our model provides a quantitative framework for unbiased what-if analyses including expected value, cost, and uncertainty.* The analyses and applications described in Chapters Three and Four demonstrate that the process outlined in Chapter Two, in addition to allowing for optimization of ATO portfolios, can also provide a quantitative basis for performing a variety of sensitivity analyses that may be desired by decisionmakers faced with uncertainties in requirements, budgets, and the outcomes of S&T projects.

Recommendations

In TAS-1 (pp. 64–65), we suggested that the Army S&T community use our process to inform their deliberations by integrating it into the existing Army S&T technology planning, review, and oversight process (DoD, 2002, pp. 4–5). In that study, we further suggested that the Army test the RAND process over a two-year period,[2] with the first year needed for establishing a baseline. For the requirements, the Army could use the latest capability needs requirements provided by TRADOC/ARCIC.

The key stakeholders involved would be the S&T project managers, the Deputy Assistant Secretary of the Army for Research and Technology [DAS (R&T)], and the Warfighting Technical Council (WTC).

In the first year, the S&T project managers and the DAS (R&T) would work together to establish an initial baseline for both performance and cost. This process would follow several steps:

- S&T project managers would estimate (a) the contribution that each S&T project that leads to a fielded system will make to Army capabilities and (b) the lifecycle cost of the system derived from each S&T project.[3]
- They would then provide these data to the DAS (R&T).
- At the same time, the DAS (R&T) would assemble data on the S&T cost to complete each project.
- Taking the estimates provided by the S&T project managers, the DAS (R&T) would use a Delphi (or other) method to gather expert opinion on whether the estimates are appropriate, too high, or too low.
- The DAS (R&T) would then apply RAND's process to inform S&T project managers of his or her office's current assessment of their projects. The process

[2] Of course, the period can be greatly shortened if the Army commits greater resources to the test.

[3] The recent work led by Martha Roper at the Office of the Deputy Assistant Secretary of the Army for Cost and Economics in assembling lifecycle cost and performance of the Army's and the other services' programs can serve as the historical basis for this cost estimation. See U.S. Army, undated.

includes RAND's LPM and simulation (as described in Chapters Two and Four) and would provide outputs such as those shown in Figures 4.1–4.4 and Tables 4.1 and 4.2.
- As the final step in the first fiscal year, the DAS (R&T) would invite the S&T project managers to improve their baseline estimates.

In the second fiscal year, estimates would be refined and decisions implemented. WTC would then join the S&T managers and the DAS (R&T) in the process:

- In the first six months of the second year, the S&T project managers would provide their adjusted estimates and justify to WTC any differences from their baseline estimates.
- In the second six months of the year, the DAS (R&T) would reapply RAND's process, with adjusted estimates where relevant, and provide the revised outputs to WTC.
- Toward the end of the year, WTC would make decisions based on the revised outputs and take any necessary corrective actions with respect to the S&T portfolio.

Once the RAND process is tested and validated, it can be integrated into the Army's S&T annual process as described in TAS-1. We anticipate that the exchanges between WTC and ATO personnel would improve the capability and lifecycle cost estimates, allowing the goal of fielding the most cost-effective systems to be factored into the design and development of the S&T programs. These exchanges would alert WTC and ATO managers to potential problems and allow them to take corrective actions in a timely manner.

In TAS-1, we also recommended three actions for designing, funding, and managing the Army's S&T portfolio so as to meet all of the Army's FOC requirements at the lowest total remaining lifecycle cost. We reiterate these recommendations below, updated to reflect the results described in this monograph.

1. Establish a pilot program to test the practicality and usefulness of the iterative procedure described above for better estimating an ATO's contributions to individual FOC requirements and its derived system's lifecycle cost. This pilot program, if successful, will be an important step toward the goal of meeting future capabilities at the lowest cost.
2. Set up a pilot program to carry out the process demonstrated in this monograph. The pilot program can be based on existing ATOs and should include all the applications discussed in Chapters Three and Four, implementing the findings described above in this chapter. We recommend the use of a Delphi method structured according to the surrogate Delphi method described in Appendix B for estimating the lifecycle cost of ATOs.

3. Since the process developed here is applicable to the portfolio management of Army programs at stages other than the S&T stage, apply the approach here to the selection and management of programs in the EMD stage.[4] Other services and DoD may wish to try this approach as well. Most important, the process should be applied to capability portfolio management as required by the Vice Chief of Staff of the Army and by DoD.

We believe that not only the Army, but also the other services and DoD as a whole, need to consider lifecycle cost at an early stage in the development of weapon and other systems in order to allow adjustments, where necessary, to achieve affordable systems that meet all individual capability requirements. By providing a method to treat uncertainties, we hope that this study can empower those who are still hesitant to make and use uncertain estimates at the S&T stage to take on the challenge of addressing, as opposed to ignoring, uncertainties.

The changes in DoD's acquisition management system in 2008 discussed in Chapter One, as well as the mandate to use capability portfolio management, should reinforce the need to consider costs and uncertainties with a portfolio process throughout the acquisition stages in order for the Army to have affordable and effective systems even under budgetary constraints.

---

[4] Under a follow-on study, we are currently applying the method and model to systems at the EMD stage and other near-term solutions to current force capability gaps defined by TRADOC.

APPENDIX A

Estimates of Expected Values

This appendix presents the results of the expected value estimates based on the gap-space coverage matrix of Figures 3.1 through 3.8. To perform these estimates, the study team performed the following analyses for each matrix element in each figure:[1]

- Determine the situations and categories to which the gap applies.
- Determine the situations and categories that the ATO addresses.
- Estimate the coverage score of the ATO for that gap as the fraction of the situations times the fraction of the categories, to which the gap applies, that the ATO addresses.
- Multiply by one-half (the assumed technical potential of each ATO, as described in Chapter Two) to convert the coverage score to the expected value of the ATO for that gap.

[1] We emphasize that the purpose of these EV estimates is to demonstrate our method, not to make decisions on ATOs. An Excel spreadsheet that identifies the gaps and shows the study team's assignments of gaps and ATOs to situations and FOC categories is available with the approval of the sponsor of this study. If interested, please see the RAND contact information in the Preface of this monograph.

Figure A.1
ATOs 1–38 Individual Gap EVs for FOC 1 Battle Command and FOC 2 Battlespace Awareness

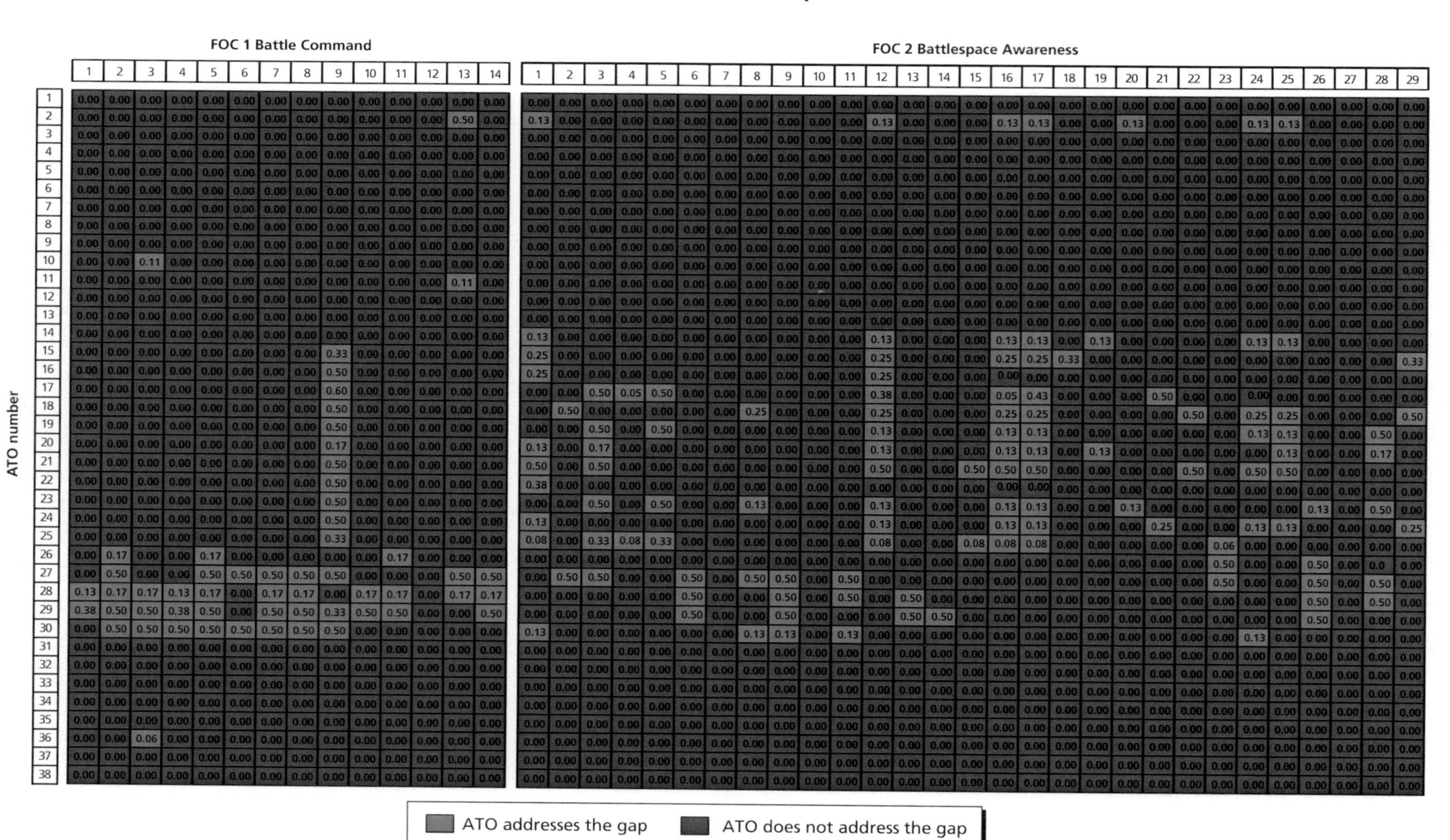

RAND *MG979-A.1*

Figure A.2
ATOs 1–38 Individual Gap EVs for FOC 3 Mounted-Dismounted Maneuver, FOC 4 Air Maneuver, FOC 5 Lethality, and FOC 6 Maneuver Support

ATO number	FOC 3 Mounted-Dismounted Maneuver		FOC 4 Air Maneuver	FOC 5 Lethality		FOC 6 Maneuver Support									
	1	2	1	1	2	1	2	3	4	5	6	7	8	9	10
1	0.00	0.00	0.00	0.00	0.00	0.00	0.00	0.00	0.00	0.00	0.00	0.50	0.00	0.50	0.00
2	0.00	0.00	0.00	0.00	0.00	0.00	0.00	0.00	0.00	0.00	0.00	0.50	0.50	0.00	0.00
3	0.00	0.00	0.13	0.00	0.00	0.00	0.00	0.00	0.00	0.00	0.00	0.00	0.00	0.00	0.00
4	0.00	0.00	0.13	0.00	0.17	0.00	0.00	0.00	0.00	0.00	0.00	0.00	0.00	0.00	0.00
5	0.00	0.00	0.00	0.00	0.00	0.00	0.00	0.00	0.00	0.00	0.00	0.00	0.00	0.00	0.00
6	0.00	0.00	0.19	0.00	0.00	0.00	0.00	0.00	0.00	0.00	0.00	0.00	0.00	0.00	0.00
7	0.00	0.00	0.00	0.00	0.00	0.00	0.00	0.00	0.00	0.00	0.00	0.00	0.00	0.00	0.00
8	0.00	0.00	0.00	0.00	0.17	0.00	0.00	0.00	0.00	0.00	0.00	0.00	0.00	0.00	0.00
9	0.00	0.00	0.00	0.00	0.00	0.00	0.00	0.00	0.00	0.00	0.25	0.25	0.00	0.00	0.00
10	0.00	0.00	0.00	0.00	0.17	0.00	0.00	0.17	0.00	0.00	0.00	0.00	0.00	0.00	0.00
11	0.00	0.00	0.00	0.00	0.00	0.00	0.00	0.00	0.00	0.00	0.50	0.50	0.00	0.00	0.00
12	0.00	0.00	0.00	0.00	0.00	0.00	0.00	0.00	0.00	0.00	0.00	0.00	0.00	0.00	0.00
13	0.00	0.00	0.00	0.00	0.00	0.00	0.00	0.00	0.00	0.00	0.00	0.00	0.00	0.00	0.00
14	0.00	0.00	0.00	0.00	0.00	0.00	0.00	0.00	0.00	0.00	0.00	0.50	0.50	0.00	0.00
15	0.00	0.00	0.00	0.00	0.00	0.00	0.00	0.00	0.00	0.00	0.00	0.00	0.00	0.00	0.00
16	0.00	0.00	0.00	0.00	0.00	0.00	0.00	0.00	0.00	0.00	0.00	0.00	0.00	0.00	0.00
17	0.00	0.00	0.00	0.00	0.00	0.00	0.00	0.00	0.00	0.00	0.00	0.00	0.00	0.00	0.00
18	0.00	0.00	0.25	0.00	0.00	0.00	0.00	0.00	0.00	0.00	0.00	0.00	0.00	0.00	0.00
19	0.00	0.00	0.00	0.00	0.00	0.00	0.00	0.00	0.00	0.00	0.00	0.00	0.00	0.00	0.50
20	0.00	0.00	0.00	0.00	0.00	0.00	0.00	0.00	0.00	0.00	0.00	0.00	0.00	0.00	0.00
21	0.00	0.00	0.00	0.00	0.00	0.00	0.00	0.00	0.00	0.00	0.00	0.00	0.00	0.00	0.50
22	0.00	0.00	0.13	0.00	0.00	0.00	0.00	0.00	0.00	0.00	0.00	0.00	0.00	0.00	0.25
23	0.00	0.00	0.00	0.00	0.00	0.00	0.00	0.00	0.00	0.00	0.00	0.00	0.00	0.00	0.25
24	0.00	0.00	0.00	0.00	0.00	0.00	0.00	0.00	0.00	0.00	0.00	0.00	0.00	0.00	0.00
25	0.00	0.00	0.00	0.00	0.00	0.00	0.00	0.00	0.00	0.00	0.00	0.00	0.00	0.00	0.00
26	0.00	0.00	0.00	0.00	0.00	0.00	0.00	0.00	0.00	0.00	0.00	0.00	0.00	0.00	0.50
27	0.50	0.50	0.00	0.00	0.00	0.00	0.00	0.00	0.00	0.00	0.00	0.00	0.00	0.00	0.50
28	0.50	0.50	0.00	0.00	0.00	0.00	0.00	0.00	0.00	0.00	0.00	0.00	0.00	0.00	0.50
29	0.50	0.50	0.00	0.00	0.00	0.00	0.00	0.00	0.00	0.00	0.00	0.00	0.00	0.00	0.50
30	0.00	0.13	0.00	0.00	0.00	0.00	0.00	0.00	0.00	0.00	0.00	0.00	0.00	0.00	0.00
31	0.00	0.00	0.06	0.00	0.50	0.00	0.00	0.00	0.00	0.00	0.00	0.00	0.00	0.00	0.00
32	0.00	0.00	0.00	0.00	0.33	0.00	0.00	0.00	0.00	0.00	0.00	0.00	0.00	0.00	0.00
33	0.00	0.00	0.00	0.50	0.50	0.00	0.00	0.00	0.00	0.00	0.00	0.00	0.00	0.00	0.00
34	0.00	0.00	0.00	0.00	0.50	0.50	0.50	0.50	0.00	0.00	0.00	0.00	0.00	0.25	0.00
35	0.00	0.00	0.00	0.00	0.17	0.00	0.00	0.00	0.00	0.00	0.00	0.00	0.00	0.00	0.00
36	0.00	0.00	0.06	0.50	0.50	0.00	0.00	0.00	0.00	0.00	0.00	0.00	0.00	0.00	0.00
37	0.00	0.00	0.00	0.50	0.50	0.00	0.00	0.00	0.00	0.00	0.00	0.00	0.00	0.00	0.00
38	0.00	0.00	0.00	0.50	0.50	0.00	0.00	0.00	0.00	0.00	0.00	0.00	0.00	0.00	0.00

ATO addresses the gap (light shading) / ATO does not address the gap (dark shading)

RAND *MG979-A.2*

Figure A.3
ATOs 1–38 Individual Gap EVs for FOC 7 Protection and FOC 8 Strategic Responsiveness and Deployability

ATO number	FOC 7 Protection																								FOC 8 Strategic Responsiveness and Deployability	
	1	2	3	4	5	6	7	8	9	10	11	12	13	14	15	16	17	18	19	20	21	22	23	24	1	2
1	0.00	0.00	0.50	0.00	0.00	0.00	0.00	0.00	0.25	0.00	0.00	0.50	0.50	0.50	0.00	0.00	0.50	0.33	0.50	0.00	0.00	0.00	0.00	0.00	0.00	0.00
2	0.00	0.00	0.00	0.00	0.00	0.00	0.00	0.00	0.25	0.00	0.00	0.50	0.50	0.50	0.50	0.50	0.50	0.33	0.50	0.00	0.00	0.00	0.00	0.00	0.17	0.00
3	0.00	0.00	0.00	0.00	0.00	0.00	0.00	0.00	0.00	0.00	0.00	0.00	0.00	0.00	0.00	0.00	0.00	0.00	0.00	0.00	0.00	0.00	0.00	0.00	0.25	0.00
4	0.00	0.00	0.00	0.00	0.00	0.00	0.00	0.00	0.00	0.00	0.25	0.00	0.00	0.00	0.00	0.50	0.00	0.17	0.25	0.25	0.00	0.17	0.00	0.00	0.00	0.00
5	0.00	0.00	0.00	0.00	0.00	0.00	0.00	0.00	0.00	0.00	0.50	0.00	0.00	0.00	0.00	0.50	0.50	0.33	0.50	0.50	0.50	0.33	0.00	0.00	0.00	0.00
6	0.00	0.00	0.00	0.00	0.00	0.00	0.00	0.00	0.00	0.00	0.25	0.00	0.00	0.00	0.00	0.50	0.00	0.17	0.25	0.25	0.00	0.17	0.00	0.00	0.00	0.00
7	0.00	0.00	0.00	0.00	0.00	0.00	0.00	0.00	0.00	0.00	0.50	0.00	0.00	0.00	0.00	0.50	0.50	0.33	0.50	0.50	0.50	0.00	0.00	0.00	0.00	0.00
8	0.00	0.00	0.00	0.00	0.00	0.00	0.00	0.00	0.00	0.00	0.25	0.00	0.00	0.00	0.00	0.50	0.00	0.17	0.25	0.25	0.00	0.17	0.25	0.00	0.00	0.00
9	0.00	0.00	0.00	0.00	0.00	0.00	0.00	0.00	0.00	0.50	0.50	0.50	0.00	0.00	0.50	0.50	0.50	0.33	0.50	0.50	0.50	0.50	0.50	0.00	0.50	0.00
10	0.00	0.00	0.00	0.00	0.00	0.00	0.00	0.00	0.00	0.00	0.50	0.00	0.00	0.00	0.00	0.50	0.50	0.00	0.00	0.50	0.50	0.00	0.00	0.00	0.00	0.00
11	0.00	0.00	0.00	0.00	0.00	0.00	0.00	0.00	0.00	0.00	0.00	0.50	0.50	0.50	0.00	0.50	0.50	0.33	0.50	0.00	0.00	0.00	0.00	0.00	0.00	0.00
12	0.00	0.00	0.00	0.00	0.00	0.00	0.00	0.00	0.25	0.00	0.00	0.00	0.00	0.00	0.00	0.00	0.00	0.08	0.13	0.00	0.00	0.00	0.00	0.00	0.00	0.00
13	0.50	0.17	0.00	0.00	0.00	0.00	0.00	0.00	0.25	0.33	0.50	0.33	0.00	0.00	0.33	0.00	0.50	0.33	0.50	0.00	0.50	0.00	0.00	0.00	0.00	0.00
14	0.00	0.00	0.00	0.00	0.00	0.00	0.00	0.00	0.25	0.00	0.00	0.50	0.50	0.00	0.00	0.00	0.50	0.33	0.50	0.00	0.00	0.00	0.00	0.00	0.33	0.00
15	0.00	0.00	0.00	0.00	0.00	0.00	0.00	0.00	0.25	0.00	0.00	0.00	0.00	0.00	0.00	0.00	0.00	0.33	0.50	0.00	0.00	0.00	0.00	0.00	0.00	0.00
16	0.00	0.00	0.00	0.00	0.00	0.00	0.00	0.00	0.00	0.00	0.00	0.00	0.00	0.00	0.00	0.50	0.00	0.33	0.50	0.00	0.00	0.00	0.00	0.00	0.00	0.00
17	0.00	0.00	0.00	0.00	0.00	0.00	0.00	0.00	0.25	0.00	0.00	0.00	0.00	0.00	0.00	0.00	0.00	0.17	0.00	0.00	0.00	0.00	0.00	0.00	0.00	0.00
18	0.00	0.00	0.00	0.00	0.00	0.00	0.00	0.00	0.00	0.00	0.00	0.00	0.00	0.00	0.00	0.50	0.50	0.33	0.50	0.00	0.00	0.00	0.00	0.00	0.00	0.00
19	0.00	0.00	0.00	0.00	0.00	0.00	0.00	0.00	0.00	0.00	0.00	0.00	0.00	0.00	0.00	0.00	0.50	0.33	0.50	0.00	0.00	0.00	0.00	0.00	0.00	0.00
20	0.00	0.00	0.00	0.00	0.00	0.00	0.00	0.00	0.00	0.00	0.00	0.00	0.00	0.00	0.00	0.50	0.00	0.00	0.25	0.00	0.00	0.00	0.00	0.00	0.00	0.00
21	0.00	0.00	0.00	0.00	0.00	0.00	0.00	0.00	0.25	0.00	0.00	0.00	0.00	0.00	0.00	0.00	0.50	0.33	0.50	0.00	0.00	0.00	0.00	0.00	0.00	0.00
22	0.00	0.00	0.00	0.00	0.00	0.00	0.00	0.00	0.00	0.00	0.00	0.00	0.00	0.00	0.00	0.00	0.00	0.00	0.00	0.00	0.00	0.00	0.00	0.00	0.50	0.00
23	0.00	0.00	0.00	0.00	0.00	0.00	0.00	0.00	0.00	0.00	0.00	0.00	0.00	0.00	0.00	0.00	0.00	0.00	0.00	0.00	0.00	0.00	0.00	0.00	0.00	0.00
24	0.00	0.00	0.00	0.00	0.00	0.00	0.00	0.00	0.25	0.00	0.00	0.00	0.00	0.00	0.00	0.00	0.50	0.33	0.50	0.00	0.00	0.00	0.00	0.00	0.00	0.00
25	0.00	0.00	0.00	0.00	0.00	0.00	0.00	0.00	0.25	0.00	0.00	0.00	0.00	0.00	0.00	0.00	0.00	0.17	0.25	0.00	0.00	0.00	0.00	0.00	0.00	0.00
26	0.00	0.00	0.00	0.00	0.00	0.00	0.00	0.00	0.00	0.00	0.00	0.00	0.00	0.00	0.00	0.00	0.17	0.00	0.00	0.00	0.00	0.00	0.00	0.50	0.00	0.00
27	0.00	0.00	0.00	0.00	0.00	0.00	0.33	0.33	0.50	0.00	0.00	0.00	0.00	0.00	0.00	0.00	0.00	0.50	0.00	0.00	0.00	0.00	0.00	0.00	0.33	0.00
28	0.00	0.00	0.00	0.00	0.00	0.00	0.00	0.00	0.00	0.00	0.00	0.00	0.00	0.00	0.00	0.00	0.00	0.00	0.00	0.00	0.00	0.00	0.00	0.00	0.00	0.00
29	0.00	0.00	0.00	0.00	0.00	0.00	0.00	0.00	0.25	0.00	0.00	0.00	0.00	0.00	0.00	0.00	0.00	0.00	0.00	0.00	0.00	0.00	0.00	0.00	0.33	0.00
30	0.00	0.00	0.00	0.00	0.00	0.00	0.00	0.33	0.50	0.00	0.00	0.00	0.00	0.00	0.00	0.00	0.00	0.00	0.00	0.00	0.00	0.00	0.00	0.00	0.50	0.00
31	0.00	0.00	0.00	0.00	0.00	0.00	0.00	0.00	0.00	0.00	0.13	0.00	0.00	0.00	0.00	0.00	0.00	0.08	0.13	0.00	0.00	0.00	0.00	0.00	0.00	0.00
32	0.00	0.00	0.00	0.00	0.00	0.00	0.00	0.00	0.00	0.00	0.25	0.00	0.00	0.00	0.00	0.00	0.00	0.00	0.00	0.00	0.00	0.00	0.00	0.00	0.00	0.00
33	0.00	0.00	0.00	0.00	0.00	0.00	0.00	0.00	0.25	0.00	0.25	0.00	0.00	0.00	0.00	0.00	0.00	0.17	0.25	0.00	0.00	0.00	0.00	0.00	0.00	0.00
34	0.00	0.00	0.00	0.00	0.00	0.00	0.00	0.00	0.25	0.00	0.00	0.00	0.00	0.00	0.00	0.00	0.00	0.00	0.00	0.00	0.00	0.00	0.00	0.00	0.00	0.00
35	0.00	0.00	0.00	0.00	0.00	0.00	0.00	0.00	0.00	0.00	0.00	0.00	0.00	0.00	0.00	0.00	0.00	0.17	0.25	0.00	0.00	0.00	0.00	0.00	0.00	0.00
36	0.00	0.00	0.00	0.00	0.00	0.00	0.00	0.00	0.25	0.00	0.00	0.00	0.00	0.00	0.00	0.00	0.00	0.00	0.25	0.00	0.00	0.00	0.00	0.00	0.00	0.00
37	0.00	0.00	0.00	0.00	0.00	0.00	0.00	0.00	0.00	0.00	0.00	0.00	0.00	0.00	0.00	0.00	0.00	0.00	0.00	0.00	0.00	0.00	0.00	0.00	0.00	0.00
38	0.00	0.00	0.00	0.00	0.00	0.00	0.00	0.00	0.25	0.00	0.00	0.00	0.00	0.00	0.00	0.00	0.00	0.17	0.25	0.00	0.00	0.00	0.00	0.00	0.00	0.00

ATO addresses the gap | ATO does not address the gap

RAND *MG979-A.3*

Figure A.4
ATOs 1–38 Individual Gap EVs for FOC 9 Maneuver Sustainment; FOC 10 Training, Education, and Leadership; and FOC 11 Human Engineering

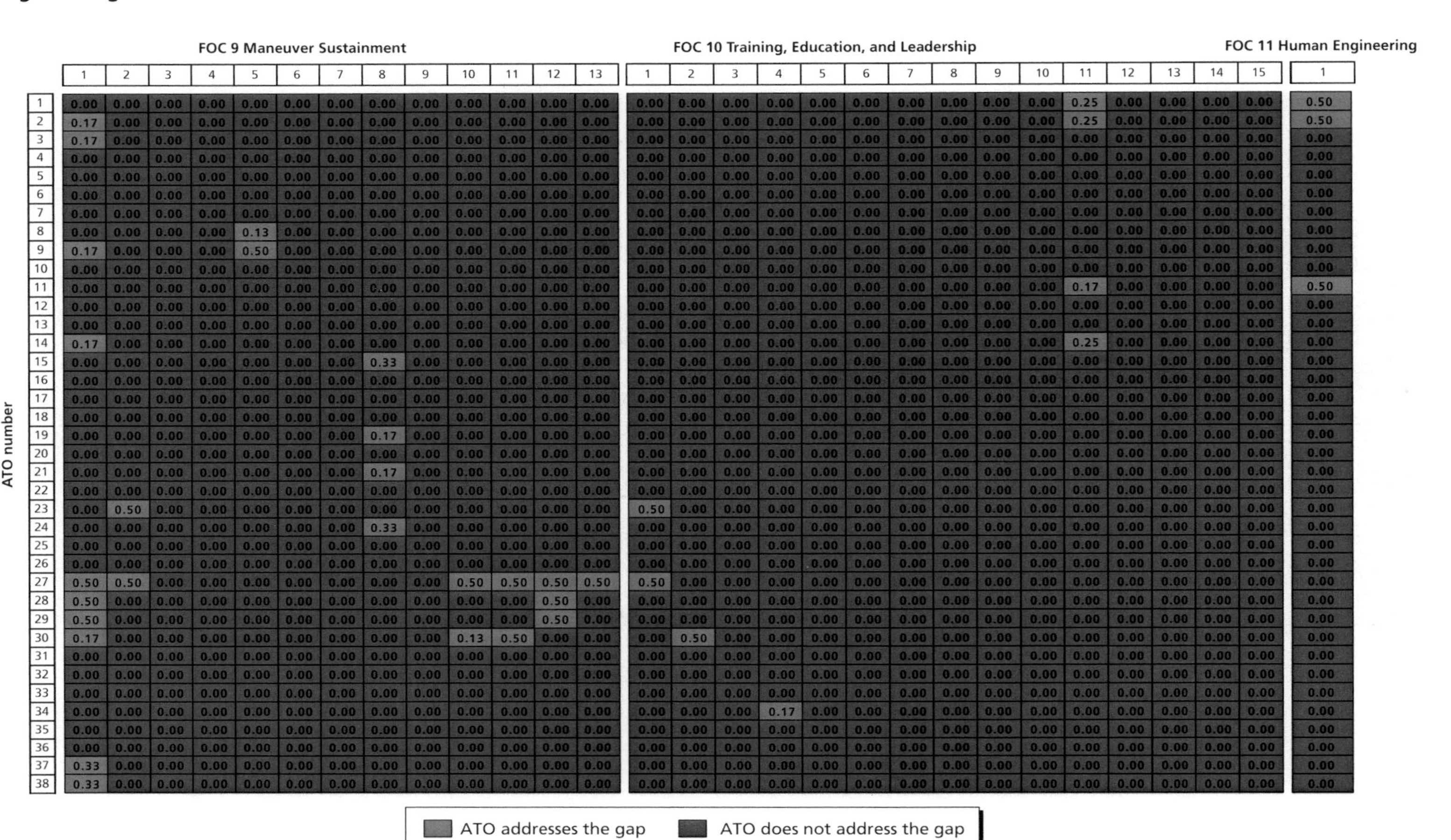

ATO number	FOC 9 Maneuver Sustainment													FOC 10 Training, Education, and Leadership															FOC 11 Human Engineering
	1	2	3	4	5	6	7	8	9	10	11	12	13	1	2	3	4	5	6	7	8	9	10	11	12	13	14	15	1
1	0.00	0.00	0.00	0.00	0.00	0.00	0.00	0.00	0.00	0.00	0.00	0.00	0.00	0.00	0.00	0.00	0.00	0.00	0.00	0.00	0.00	0.00	0.00	0.25	0.00	0.00	0.00	0.00	0.50
2	0.17	0.00	0.00	0.00	0.00	0.00	0.00	0.00	0.00	0.00	0.00	0.00	0.00	0.00	0.00	0.00	0.00	0.00	0.00	0.00	0.00	0.00	0.00	0.25	0.00	0.00	0.00	0.00	0.50
3	0.17	0.00	0.00	0.00	0.00	0.00	0.00	0.00	0.00	0.00	0.00	0.00	0.00	0.00	0.00	0.00	0.00	0.00	0.00	0.00	0.00	0.00	0.00	0.00	0.00	0.00	0.00	0.00	0.00
4	0.00	0.00	0.00	0.00	0.00	0.00	0.00	0.00	0.00	0.00	0.00	0.00	0.00	0.00	0.00	0.00	0.00	0.00	0.00	0.00	0.00	0.00	0.00	0.00	0.00	0.00	0.00	0.00	0.00
5	0.00	0.00	0.00	0.00	0.00	0.00	0.00	0.00	0.00	0.00	0.00	0.00	0.00	0.00	0.00	0.00	0.00	0.00	0.00	0.00	0.00	0.00	0.00	0.00	0.00	0.00	0.00	0.00	0.00
6	0.00	0.00	0.00	0.00	0.00	0.00	0.00	0.00	0.00	0.00	0.00	0.00	0.00	0.00	0.00	0.00	0.00	0.00	0.00	0.00	0.00	0.00	0.00	0.00	0.00	0.00	0.00	0.00	0.00
7	0.00	0.00	0.00	0.00	0.00	0.00	0.00	0.00	0.00	0.00	0.00	0.00	0.00	0.00	0.00	0.00	0.00	0.00	0.00	0.00	0.00	0.00	0.00	0.00	0.00	0.00	0.00	0.00	0.00
8	0.00	0.00	0.00	0.00	0.13	0.00	0.00	0.00	0.00	0.00	0.00	0.00	0.00	0.00	0.00	0.00	0.00	0.00	0.00	0.00	0.00	0.00	0.00	0.00	0.00	0.00	0.00	0.00	0.00
9	0.17	0.00	0.00	0.00	0.50	0.00	0.00	0.00	0.00	0.00	0.00	0.00	0.00	0.00	0.00	0.00	0.00	0.00	0.00	0.00	0.00	0.00	0.00	0.00	0.00	0.00	0.00	0.00	0.00
10	0.00	0.00	0.00	0.00	0.00	0.00	0.00	0.00	0.00	0.00	0.00	0.00	0.00	0.00	0.00	0.00	0.00	0.00	0.00	0.00	0.00	0.00	0.00	0.00	0.00	0.00	0.00	0.00	0.00
11	0.00	0.00	0.00	0.00	0.00	0.00	0.00	0.00	0.00	0.00	0.00	0.00	0.00	0.00	0.00	0.00	0.00	0.00	0.00	0.00	0.00	0.00	0.00	0.17	0.00	0.00	0.00	0.00	0.50
12	0.00	0.00	0.00	0.00	0.00	0.00	0.00	0.00	0.00	0.00	0.00	0.00	0.00	0.00	0.00	0.00	0.00	0.00	0.00	0.00	0.00	0.00	0.00	0.00	0.00	0.00	0.00	0.00	0.00
13	0.00	0.00	0.00	0.00	0.00	0.00	0.00	0.00	0.00	0.00	0.00	0.00	0.00	0.00	0.00	0.00	0.00	0.00	0.00	0.00	0.00	0.00	0.00	0.00	0.00	0.00	0.00	0.00	0.00
14	0.17	0.00	0.00	0.00	0.00	0.00	0.00	0.00	0.00	0.00	0.00	0.00	0.00	0.00	0.00	0.00	0.00	0.00	0.00	0.00	0.00	0.00	0.00	0.25	0.00	0.00	0.00	0.00	0.00
15	0.00	0.00	0.00	0.00	0.00	0.00	0.00	0.33	0.00	0.00	0.00	0.00	0.00	0.00	0.00	0.00	0.00	0.00	0.00	0.00	0.00	0.00	0.00	0.00	0.00	0.00	0.00	0.00	0.00
16	0.00	0.00	0.00	0.00	0.00	0.00	0.00	0.00	0.00	0.00	0.00	0.00	0.00	0.00	0.00	0.00	0.00	0.00	0.00	0.00	0.00	0.00	0.00	0.00	0.00	0.00	0.00	0.00	0.00
17	0.00	0.00	0.00	0.00	0.00	0.00	0.00	0.00	0.00	0.00	0.00	0.00	0.00	0.00	0.00	0.00	0.00	0.00	0.00	0.00	0.00	0.00	0.00	0.00	0.00	0.00	0.00	0.00	0.00
18	0.00	0.00	0.00	0.00	0.00	0.00	0.00	0.00	0.00	0.00	0.00	0.00	0.00	0.00	0.00	0.00	0.00	0.00	0.00	0.00	0.00	0.00	0.00	0.00	0.00	0.00	0.00	0.00	0.00
19	0.00	0.00	0.00	0.00	0.00	0.00	0.00	0.17	0.00	0.00	0.00	0.00	0.00	0.00	0.00	0.00	0.00	0.00	0.00	0.00	0.00	0.00	0.00	0.00	0.00	0.00	0.00	0.00	0.00
20	0.00	0.00	0.00	0.00	0.00	0.00	0.00	0.00	0.00	0.00	0.00	0.00	0.00	0.00	0.00	0.00	0.00	0.00	0.00	0.00	0.00	0.00	0.00	0.00	0.00	0.00	0.00	0.00	0.00
21	0.00	0.00	0.00	0.00	0.00	0.00	0.00	0.17	0.00	0.00	0.00	0.00	0.00	0.00	0.00	0.00	0.00	0.00	0.00	0.00	0.00	0.00	0.00	0.00	0.00	0.00	0.00	0.00	0.00
22	0.00	0.00	0.00	0.00	0.00	0.00	0.00	0.00	0.00	0.00	0.00	0.00	0.00	0.00	0.00	0.00	0.00	0.00	0.00	0.00	0.00	0.00	0.00	0.00	0.00	0.00	0.00	0.00	0.00
23	0.00	0.50	0.00	0.00	0.00	0.00	0.00	0.00	0.00	0.00	0.00	0.00	0.00	0.50	0.00	0.00	0.00	0.00	0.00	0.00	0.00	0.00	0.00	0.00	0.00	0.00	0.00	0.00	0.00
24	0.00	0.00	0.00	0.00	0.00	0.00	0.00	0.33	0.00	0.00	0.00	0.00	0.00	0.00	0.00	0.00	0.00	0.00	0.00	0.00	0.00	0.00	0.00	0.00	0.00	0.00	0.00	0.00	0.00
25	0.00	0.00	0.00	0.00	0.00	0.00	0.00	0.00	0.00	0.00	0.00	0.00	0.00	0.00	0.00	0.00	0.00	0.00	0.00	0.00	0.00	0.00	0.00	0.00	0.00	0.00	0.00	0.00	0.00
26	0.00	0.00	0.00	0.00	0.00	0.00	0.00	0.00	0.00	0.00	0.00	0.00	0.00	0.00	0.00	0.00	0.00	0.00	0.00	0.00	0.00	0.00	0.00	0.00	0.00	0.00	0.00	0.00	0.00
27	0.50	0.50	0.00	0.00	0.00	0.00	0.00	0.00	0.00	0.50	0.50	0.50	0.50	0.50	0.00	0.00	0.00	0.00	0.00	0.00	0.00	0.00	0.00	0.00	0.00	0.00	0.00	0.00	0.00
28	0.50	0.00	0.00	0.00	0.00	0.00	0.00	0.00	0.00	0.00	0.00	0.50	0.00	0.00	0.00	0.00	0.00	0.00	0.00	0.00	0.00	0.00	0.00	0.00	0.00	0.00	0.00	0.00	0.00
29	0.50	0.00	0.00	0.00	0.00	0.00	0.00	0.00	0.00	0.00	0.00	0.50	0.00	0.00	0.00	0.00	0.00	0.00	0.00	0.00	0.00	0.00	0.00	0.00	0.00	0.00	0.00	0.00	0.00
30	0.17	0.00	0.00	0.00	0.00	0.00	0.00	0.00	0.00	0.13	0.50	0.00	0.00	0.00	0.50	0.00	0.00	0.00	0.00	0.00	0.00	0.00	0.00	0.00	0.00	0.00	0.00	0.00	0.00
31	0.00	0.00	0.00	0.00	0.00	0.00	0.00	0.00	0.00	0.00	0.00	0.00	0.00	0.00	0.00	0.00	0.00	0.00	0.00	0.00	0.00	0.00	0.00	0.00	0.00	0.00	0.00	0.00	0.00
32	0.00	0.00	0.00	0.00	0.00	0.00	0.00	0.00	0.00	0.00	0.00	0.00	0.00	0.00	0.00	0.00	0.00	0.00	0.00	0.00	0.00	0.00	0.00	0.00	0.00	0.00	0.00	0.00	0.00
33	0.00	0.00	0.00	0.00	0.00	0.00	0.00	0.00	0.00	0.00	0.00	0.00	0.00	0.00	0.00	0.00	0.00	0.00	0.00	0.00	0.00	0.00	0.00	0.00	0.00	0.00	0.00	0.00	0.00
34	0.00	0.00	0.00	0.00	0.00	0.00	0.00	0.00	0.00	0.00	0.00	0.00	0.00	0.00	0.00	0.00	0.17	0.00	0.00	0.00	0.00	0.00	0.00	0.00	0.00	0.00	0.00	0.00	0.00
35	0.00	0.00	0.00	0.00	0.00	0.00	0.00	0.00	0.00	0.00	0.00	0.00	0.00	0.00	0.00	0.00	0.00	0.00	0.00	0.00	0.00	0.00	0.00	0.00	0.00	0.00	0.00	0.00	0.00
36	0.00	0.00	0.00	0.00	0.00	0.00	0.00	0.00	0.00	0.00	0.00	0.00	0.00	0.00	0.00	0.00	0.00	0.00	0.00	0.00	0.00	0.00	0.00	0.00	0.00	0.00	0.00	0.00	0.00
37	0.33	0.00	0.00	0.00	0.00	0.00	0.00	0.00	0.00	0.00	0.00	0.00	0.00	0.00	0.00	0.00	0.00	0.00	0.00	0.00	0.00	0.00	0.00	0.00	0.00	0.00	0.00	0.00	0.00
38	0.33	0.00	0.00	0.00	0.00	0.00	0.00	0.00	0.00	0.00	0.00	0.00	0.00	0.00	0.00	0.00	0.00	0.00	0.00	0.00	0.00	0.00	0.00	0.00	0.00	0.00	0.00	0.00	0.00

RAND *MG979-A.4*

Figure A.5
ATOs 39–75 Individual Gap EVs for FOC 1 Battle Command and FOC 2 Battlespace Awareness

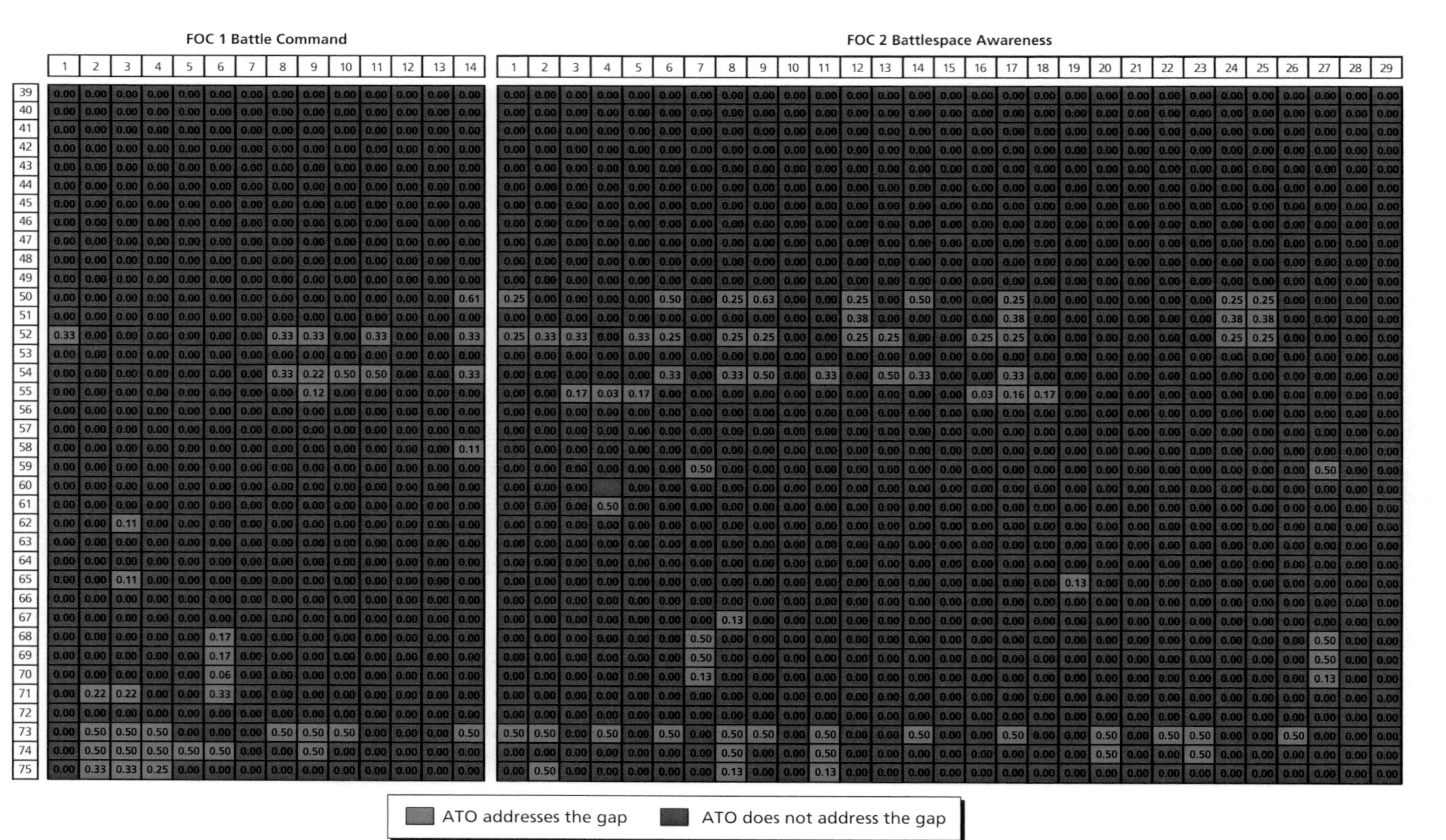

RAND *MG979-A.5*

Figure A.6
ATOs 39–75 Individual Gap EVs for FOC 3 Mounted-Dismounted Maneuver, FOC 4 Air Maneuver, FOC 5 Lethality, and FOC 6 Maneuver Support

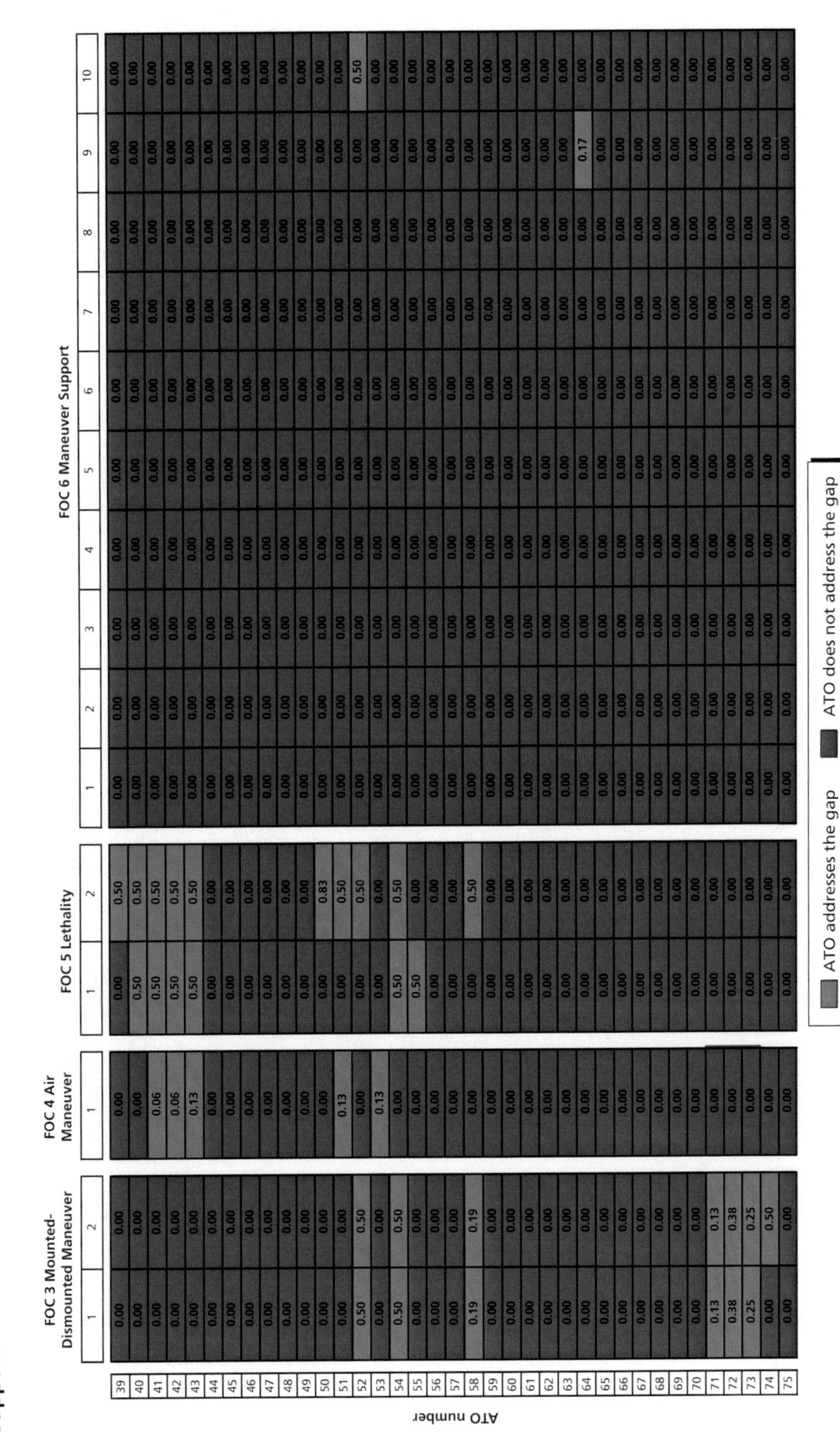

ATO number	FOC 3 Mounted-Dismounted Maneuver		FOC 4 Air Maneuver	FOC 5 Lethality		FOC 6 Maneuver Support									
	1	2	1	1	2	1	2	3	4	5	6	7	8	9	10
39	0.00	0.00	0.00	0.00	0.50	0.00	0.00	0.00	0.00	0.00	0.00	0.00	0.00	0.00	0.00
40	0.00	0.00	0.00	0.50	0.50	0.00	0.00	0.00	0.00	0.00	0.00	0.00	0.00	0.00	0.00
41	0.00	0.00	0.06	0.50	0.50	0.00	0.00	0.00	0.00	0.00	0.00	0.00	0.00	0.00	0.00
42	0.00	0.00	0.06	0.50	0.50	0.00	0.00	0.00	0.00	0.00	0.00	0.00	0.00	0.00	0.00
43	0.00	0.00	0.13	0.50	0.50	0.00	0.00	0.00	0.00	0.00	0.00	0.00	0.00	0.00	0.00
44	0.00	0.00	0.00	0.00	0.00	0.00	0.00	0.00	0.00	0.00	0.00	0.00	0.00	0.00	0.00
45	0.00	0.00	0.00	0.00	0.00	0.00	0.00	0.00	0.00	0.00	0.00	0.00	0.00	0.00	0.00
46	0.00	0.00	0.00	0.00	0.00	0.00	0.00	0.00	0.00	0.00	0.00	0.00	0.00	0.00	0.00
47	0.00	0.00	0.00	0.00	0.00	0.00	0.00	0.00	0.00	0.00	0.00	0.00	0.00	0.00	0.00
48	0.00	0.00	0.00	0.00	0.00	0.00	0.00	0.00	0.00	0.00	0.00	0.00	0.00	0.00	0.00
49	0.00	0.00	0.00	0.00	0.00	0.00	0.00	0.00	0.00	0.00	0.00	0.00	0.00	0.00	0.00
50	0.00	0.00	0.00	0.00	0.83	0.00	0.00	0.00	0.00	0.00	0.00	0.00	0.00	0.00	0.00
51	0.00	0.00	0.13	0.00	0.50	0.00	0.00	0.00	0.00	0.00	0.00	0.00	0.00	0.00	0.00
52	0.50	0.50	0.00	0.00	0.50	0.00	0.00	0.00	0.00	0.00	0.00	0.00	0.00	0.00	0.50
53	0.00	0.00	0.13	0.00	0.00	0.00	0.00	0.00	0.00	0.00	0.00	0.00	0.00	0.00	0.00
54	0.50	0.50	0.00	0.50	0.50	0.00	0.00	0.00	0.00	0.00	0.00	0.00	0.00	0.00	0.00
55	0.00	0.00	0.00	0.50	0.00	0.00	0.00	0.00	0.00	0.00	0.00	0.00	0.00	0.00	0.00
56	0.00	0.00	0.00	0.00	0.00	0.00	0.00	0.00	0.00	0.00	0.00	0.00	0.00	0.00	0.00
57	0.00	0.00	0.00	0.00	0.00	0.00	0.00	0.00	0.00	0.00	0.00	0.00	0.00	0.00	0.00
58	0.19	0.19	0.00	0.00	0.50	0.00	0.00	0.00	0.00	0.00	0.00	0.00	0.00	0.00	0.00
59	0.00	0.00	0.00	0.00	0.00	0.00	0.00	0.00	0.00	0.00	0.00	0.00	0.00	0.00	0.00
60	0.00	0.00	0.00	0.00	0.00	0.00	0.00	0.00	0.00	0.00	0.00	0.00	0.00	0.00	0.00
61	0.00	0.00	0.00	0.00	0.00	0.00	0.00	0.00	0.00	0.00	0.00	0.00	0.00	0.00	0.00
62	0.00	0.00	0.00	0.00	0.00	0.00	0.00	0.00	0.00	0.00	0.00	0.00	0.00	0.00	0.00
63	0.00	0.00	0.00	0.00	0.00	0.00	0.00	0.00	0.00	0.00	0.00	0.00	0.00	0.00	0.00
64	0.00	0.00	0.00	0.00	0.00	0.00	0.00	0.00	0.00	0.00	0.00	0.00	0.00	0.17	0.00
65	0.00	0.00	0.00	0.00	0.00	0.00	0.00	0.00	0.00	0.00	0.00	0.00	0.00	0.00	0.00
66	0.00	0.00	0.00	0.00	0.00	0.00	0.00	0.00	0.00	0.00	0.00	0.00	0.00	0.00	0.00
67	0.00	0.00	0.00	0.00	0.00	0.00	0.00	0.00	0.00	0.00	0.00	0.00	0.00	0.00	0.00
68	0.00	0.00	0.00	0.00	0.00	0.00	0.00	0.00	0.00	0.00	0.00	0.00	0.00	0.00	0.00
69	0.00	0.00	0.00	0.00	0.00	0.00	0.00	0.00	0.00	0.00	0.00	0.00	0.00	0.00	0.00
70	0.00	0.00	0.00	0.00	0.00	0.00	0.00	0.00	0.00	0.00	0.00	0.00	0.00	0.00	0.00
71	0.13	0.13	0.00	0.00	0.00	0.00	0.00	0.00	0.00	0.00	0.00	0.00	0.00	0.00	0.00
72	0.38	0.38	0.00	0.00	0.00	0.00	0.00	0.00	0.00	0.00	0.00	0.00	0.00	0.00	0.00
73	0.25	0.25	0.00	0.00	0.00	0.00	0.00	0.00	0.00	0.00	0.00	0.00	0.00	0.00	0.00
74	0.00	0.50	0.00	0.00	0.00	0.00	0.00	0.00	0.00	0.00	0.00	0.00	0.00	0.00	0.00
75	0.00	0.00	0.00	0.00	0.00	0.00	0.00	0.00	0.00	0.00	0.00	0.00	0.00	0.00	0.00

RAND *MG979-A.6*

Figure A.7
ATOs 39–75 Individual Gap EVs for FOC 7 Protection and FOC 8 Strategic Responsiveness and Deployability

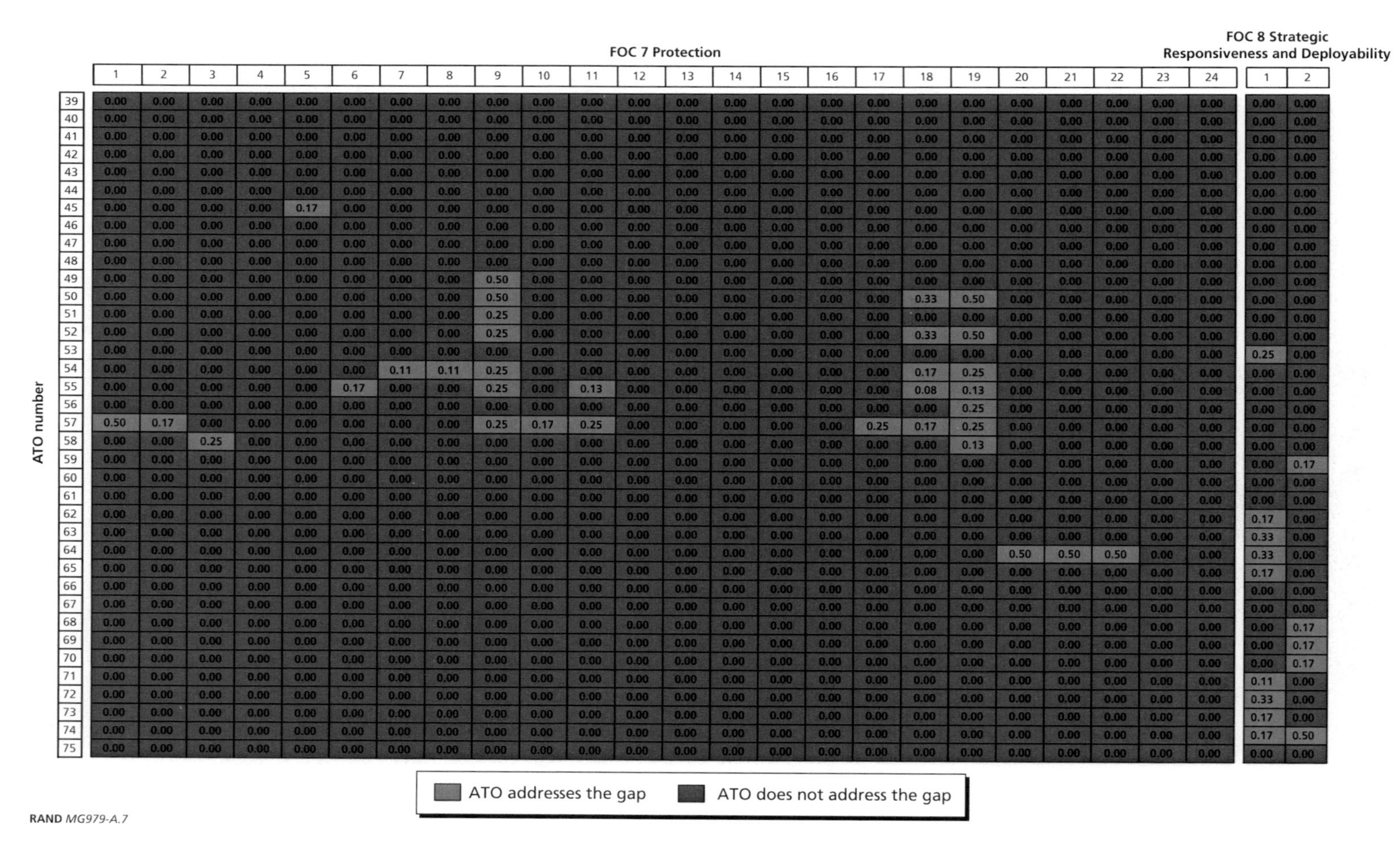

ATO number	FOC 7 Protection																								FOC 8 Strategic Responsiveness and Deployability	
	1	2	3	4	5	6	7	8	9	10	11	12	13	14	15	16	17	18	19	20	21	22	23	24	1	2
39	0.00	0.00	0.00	0.00	0.00	0.00	0.00	0.00	0.00	0.00	0.00	0.00	0.00	0.00	0.00	0.00	0.00	0.00	0.00	0.00	0.00	0.00	0.00	0.00	0.00	0.00
40	0.00	0.00	0.00	0.00	0.00	0.00	0.00	0.00	0.00	0.00	0.00	0.00	0.00	0.00	0.00	0.00	0.00	0.00	0.00	0.00	0.00	0.00	0.00	0.00	0.00	0.00
41	0.00	0.00	0.00	0.00	0.00	0.00	0.00	0.00	0.00	0.00	0.00	0.00	0.00	0.00	0.00	0.00	0.00	0.00	0.00	0.00	0.00	0.00	0.00	0.00	0.00	0.00
42	0.00	0.00	0.00	0.00	0.00	0.00	0.00	0.00	0.00	0.00	0.00	0.00	0.00	0.00	0.00	0.00	0.00	0.00	0.00	0.00	0.00	0.00	0.00	0.00	0.00	0.00
43	0.00	0.00	0.00	0.00	0.00	0.00	0.00	0.00	0.00	0.00	0.00	0.00	0.00	0.00	0.00	0.00	0.00	0.00	0.00	0.00	0.00	0.00	0.00	0.00	0.00	0.00
44	0.00	0.00	0.00	0.00	0.00	0.00	0.00	0.00	0.00	0.00	0.00	0.00	0.00	0.00	0.00	0.00	0.00	0.00	0.00	0.00	0.00	0.00	0.00	0.00	0.00	0.00
45	0.00	0.00	0.00	0.00	0.17	0.00	0.00	0.00	0.00	0.00	0.00	0.00	0.00	0.00	0.00	0.00	0.00	0.00	0.00	0.00	0.00	0.00	0.00	0.00	0.00	0.00
46	0.00	0.00	0.00	0.00	0.00	0.00	0.00	0.00	0.00	0.00	0.00	0.00	0.00	0.00	0.00	0.00	0.00	0.00	0.00	0.00	0.00	0.00	0.00	0.00	0.00	0.00
47	0.00	0.00	0.00	0.00	0.00	0.00	0.00	0.00	0.00	0.00	0.00	0.00	0.00	0.00	0.00	0.00	0.00	0.00	0.00	0.00	0.00	0.00	0.00	0.00	0.00	0.00
48	0.00	0.00	0.00	0.00	0.00	0.00	0.00	0.00	0.00	0.00	0.00	0.00	0.00	0.00	0.00	0.00	0.00	0.00	0.00	0.00	0.00	0.00	0.00	0.00	0.00	0.00
49	0.00	0.00	0.00	0.00	0.00	0.00	0.00	0.00	0.50	0.00	0.00	0.00	0.00	0.00	0.00	0.00	0.00	0.00	0.00	0.00	0.00	0.00	0.00	0.00	0.00	0.00
50	0.00	0.00	0.00	0.00	0.00	0.00	0.00	0.00	0.50	0.00	0.00	0.00	0.00	0.00	0.00	0.00	0.00	0.33	0.50	0.00	0.00	0.00	0.00	0.00	0.00	0.00
51	0.00	0.00	0.00	0.00	0.00	0.00	0.00	0.00	0.25	0.00	0.00	0.00	0.00	0.00	0.00	0.00	0.00	0.00	0.00	0.00	0.00	0.00	0.00	0.00	0.00	0.00
52	0.00	0.00	0.00	0.00	0.00	0.00	0.00	0.00	0.25	0.00	0.00	0.00	0.00	0.00	0.00	0.00	0.00	0.33	0.50	0.00	0.00	0.00	0.00	0.00	0.00	0.00
53	0.00	0.00	0.00	0.00	0.00	0.00	0.00	0.00	0.00	0.00	0.00	0.00	0.00	0.00	0.00	0.00	0.00	0.00	0.00	0.00	0.00	0.00	0.00	0.00	0.25	0.00
54	0.00	0.00	0.00	0.00	0.00	0.00	0.11	0.11	0.25	0.00	0.00	0.00	0.00	0.00	0.00	0.00	0.00	0.17	0.25	0.00	0.00	0.00	0.00	0.00	0.00	0.00
55	0.00	0.00	0.00	0.00	0.00	0.17	0.00	0.00	0.25	0.00	0.13	0.00	0.00	0.00	0.00	0.00	0.00	0.08	0.13	0.00	0.00	0.00	0.00	0.00	0.00	0.00
56	0.00	0.00	0.00	0.00	0.00	0.00	0.00	0.00	0.00	0.00	0.00	0.00	0.00	0.00	0.00	0.00	0.00	0.00	0.25	0.00	0.00	0.00	0.00	0.00	0.00	0.00
57	0.50	0.17	0.00	0.00	0.00	0.00	0.00	0.00	0.25	0.17	0.25	0.00	0.00	0.00	0.00	0.00	0.25	0.17	0.25	0.00	0.00	0.00	0.00	0.00	0.00	0.00
58	0.00	0.00	0.25	0.00	0.00	0.00	0.00	0.00	0.00	0.00	0.00	0.00	0.00	0.00	0.00	0.00	0.00	0.00	0.13	0.00	0.00	0.00	0.00	0.00	0.00	0.00
59	0.00	0.00	0.00	0.00	0.00	0.00	0.00	0.00	0.00	0.00	0.00	0.00	0.00	0.00	0.00	0.00	0.00	0.00	0.00	0.00	0.00	0.00	0.00	0.00	0.00	0.17
60	0.00	0.00	0.00	0.00	0.00	0.00	0.00	0.00	0.00	0.00	0.00	0.00	0.00	0.00	0.00	0.00	0.00	0.00	0.00	0.00	0.00	0.00	0.00	0.00	0.00	0.00
61	0.00	0.00	0.00	0.00	0.00	0.00	0.00	0.00	0.00	0.00	0.00	0.00	0.00	0.00	0.00	0.00	0.00	0.00	0.00	0.00	0.00	0.00	0.00	0.00	0.00	0.00
62	0.00	0.00	0.00	0.00	0.00	0.00	0.00	0.00	0.00	0.00	0.00	0.00	0.00	0.00	0.00	0.00	0.00	0.00	0.00	0.00	0.00	0.00	0.00	0.00	0.17	0.00
63	0.00	0.00	0.00	0.00	0.00	0.00	0.00	0.00	0.00	0.00	0.00	0.00	0.00	0.00	0.00	0.00	0.00	0.00	0.00	0.00	0.00	0.00	0.00	0.00	0.33	0.00
64	0.00	0.00	0.00	0.00	0.00	0.00	0.00	0.00	0.00	0.00	0.00	0.00	0.00	0.00	0.00	0.00	0.00	0.00	0.00	0.50	0.50	0.50	0.00	0.00	0.33	0.00
65	0.00	0.00	0.00	0.00	0.00	0.00	0.00	0.00	0.00	0.00	0.00	0.00	0.00	0.00	0.00	0.00	0.00	0.00	0.00	0.00	0.00	0.00	0.00	0.00	0.17	0.00
66	0.00	0.00	0.00	0.00	0.00	0.00	0.00	0.00	0.00	0.00	0.00	0.00	0.00	0.00	0.00	0.00	0.00	0.00	0.00	0.00	0.00	0.00	0.00	0.00	0.00	0.00
67	0.00	0.00	0.00	0.00	0.00	0.00	0.00	0.00	0.00	0.00	0.00	0.00	0.00	0.00	0.00	0.00	0.00	0.00	0.00	0.00	0.00	0.00	0.00	0.00	0.00	0.00
68	0.00	0.00	0.00	0.00	0.00	0.00	0.00	0.00	0.00	0.00	0.00	0.00	0.00	0.00	0.00	0.00	0.00	0.00	0.00	0.00	0.00	0.00	0.00	0.00	0.00	0.17
69	0.00	0.00	0.00	0.00	0.00	0.00	0.00	0.00	0.00	0.00	0.00	0.00	0.00	0.00	0.00	0.00	0.00	0.00	0.00	0.00	0.00	0.00	0.00	0.00	0.00	0.17
70	0.00	0.00	0.00	0.00	0.00	0.00	0.00	0.00	0.00	0.00	0.00	0.00	0.00	0.00	0.00	0.00	0.00	0.00	0.00	0.00	0.00	0.00	0.00	0.00	0.00	0.17
71	0.00	0.00	0.00	0.00	0.00	0.00	0.00	0.00	0.00	0.00	0.00	0.00	0.00	0.00	0.00	0.00	0.00	0.00	0.00	0.00	0.00	0.00	0.00	0.00	0.11	0.00
72	0.00	0.00	0.00	0.00	0.00	0.00	0.00	0.00	0.00	0.00	0.00	0.00	0.00	0.00	0.00	0.00	0.00	0.00	0.00	0.00	0.00	0.00	0.00	0.00	0.33	0.00
73	0.00	0.00	0.00	0.00	0.00	0.00	0.00	0.00	0.00	0.00	0.00	0.00	0.00	0.00	0.00	0.00	0.00	0.00	0.00	0.00	0.00	0.00	0.00	0.00	0.17	0.00
74	0.00	0.00	0.00	0.00	0.00	0.00	0.00	0.00	0.00	0.00	0.00	0.00	0.00	0.00	0.00	0.00	0.00	0.00	0.00	0.00	0.00	0.00	0.00	0.00	0.17	0.50
75	0.00	0.00	0.00	0.00	0.00	0.00	0.00	0.00	0.00	0.00	0.00	0.00	0.00	0.00	0.00	0.00	0.00	0.00	0.00	0.00	0.00	0.00	0.00	0.00	0.00	0.00

RAND *MG979-A.7*

Figure A.8
ATOs 39–75 Individual Gap EVs for FOC 9 Maneuver Sustainment; FOC 10 Training, Education, and Leadership; and FOC 11 Human Engineering

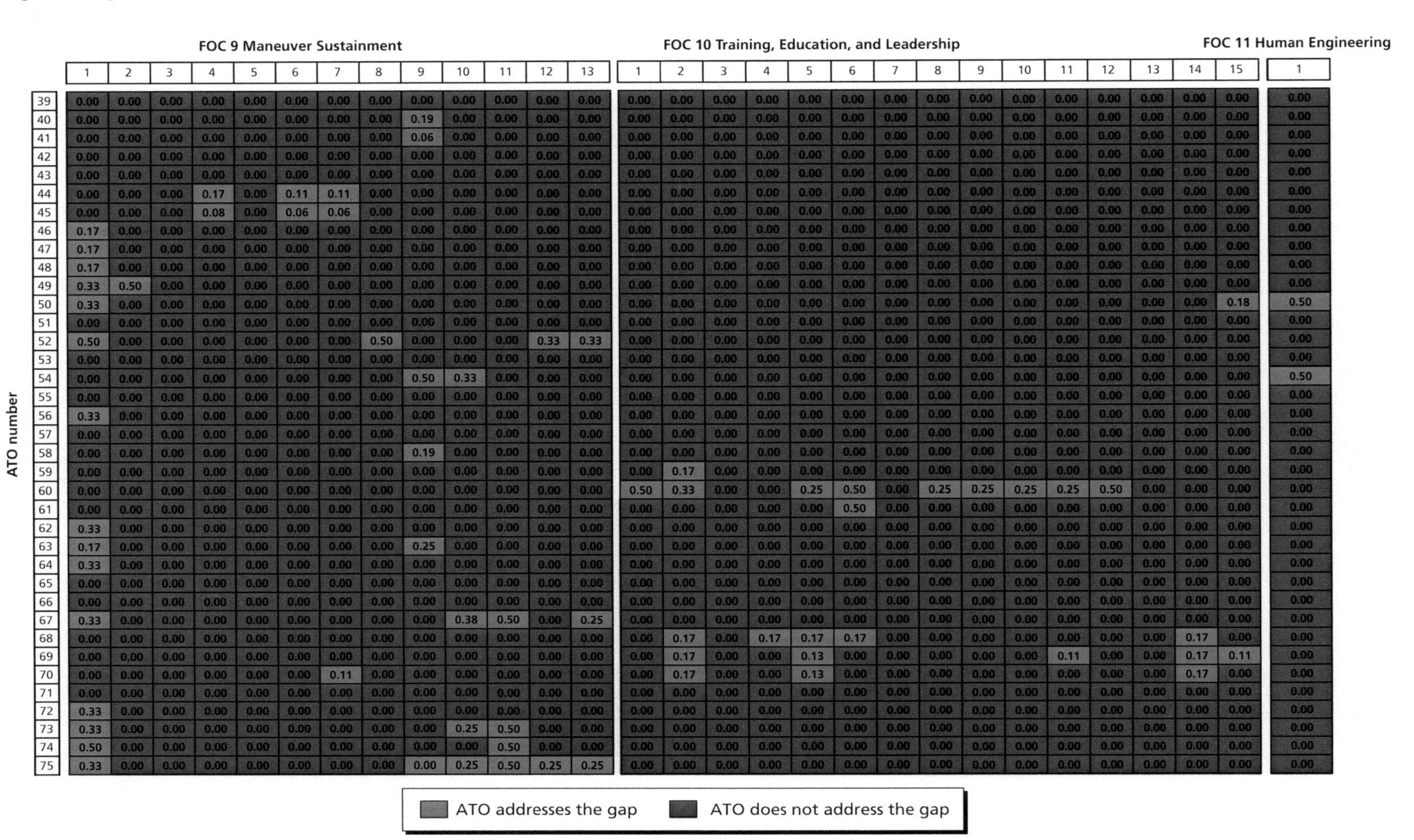

ATO number	FOC 9 Maneuver Sustainment													FOC 10 Training, Education, and Leadership															FOC 11 Human Engineering
	1	2	3	4	5	6	7	8	9	10	11	12	13	1	2	3	4	5	6	7	8	9	10	11	12	13	14	15	1
39	0.00	0.00	0.00	0.00	0.00	0.00	0.00	0.00	0.00	0.00	0.00	0.00	0.00	0.00	0.00	0.00	0.00	0.00	0.00	0.00	0.00	0.00	0.00	0.00	0.00	0.00	0.00	0.00	0.00
40	0.00	0.00	0.00	0.00	0.00	0.00	0.00	0.00	0.19	0.00	0.00	0.00	0.00	0.00	0.00	0.00	0.00	0.00	0.00	0.00	0.00	0.00	0.00	0.00	0.00	0.00	0.00	0.00	0.00
41	0.00	0.00	0.00	0.00	0.00	0.00	0.00	0.00	0.06	0.00	0.00	0.00	0.00	0.00	0.00	0.00	0.00	0.00	0.00	0.00	0.00	0.00	0.00	0.00	0.00	0.00	0.00	0.00	0.00
42	0.00	0.00	0.00	0.00	0.00	0.00	0.00	0.00	0.00	0.00	0.00	0.00	0.00	0.00	0.00	0.00	0.00	0.00	0.00	0.00	0.00	0.00	0.00	0.00	0.00	0.00	0.00	0.00	0.00
43	0.00	0.00	0.00	0.00	0.00	0.00	0.00	0.00	0.00	0.00	0.00	0.00	0.00	0.00	0.00	0.00	0.00	0.00	0.00	0.00	0.00	0.00	0.00	0.00	0.00	0.00	0.00	0.00	0.00
44	0.00	0.00	0.00	0.17	0.00	0.11	0.11	0.00	0.00	0.00	0.00	0.00	0.00	0.00	0.00	0.00	0.00	0.00	0.00	0.00	0.00	0.00	0.00	0.00	0.00	0.00	0.00	0.00	0.00
45	0.00	0.00	0.00	0.08	0.00	0.06	0.06	0.00	0.00	0.00	0.00	0.00	0.00	0.00	0.00	0.00	0.00	0.00	0.00	0.00	0.00	0.00	0.00	0.00	0.00	0.00	0.00	0.00	0.00
46	0.17	0.00	0.00	0.00	0.00	0.00	0.00	0.00	0.00	0.00	0.00	0.00	0.00	0.00	0.00	0.00	0.00	0.00	0.00	0.00	0.00	0.00	0.00	0.00	0.00	0.00	0.00	0.00	0.00
47	0.17	0.00	0.00	0.00	0.00	0.00	0.00	0.00	0.00	0.00	0.00	0.00	0.00	0.00	0.00	0.00	0.00	0.00	0.00	0.00	0.00	0.00	0.00	0.00	0.00	0.00	0.00	0.00	0.00
48	0.17	0.00	0.00	0.00	0.00	0.00	0.00	0.00	0.00	0.00	0.00	0.00	0.00	0.00	0.00	0.00	0.00	0.00	0.00	0.00	0.00	0.00	0.00	0.00	0.00	0.00	0.00	0.00	0.00
49	0.33	0.50	0.00	0.00	0.00	0.00	0.00	0.00	0.00	0.00	0.00	0.00	0.00	0.00	0.00	0.00	0.00	0.00	0.00	0.00	0.00	0.00	0.00	0.00	0.00	0.00	0.00	0.00	0.00
50	0.33	0.00	0.00	0.00	0.00	0.00	0.00	0.00	0.00	0.00	0.00	0.00	0.00	0.00	0.00	0.00	0.00	0.00	0.00	0.00	0.00	0.00	0.00	0.00	0.00	0.00	0.00	0.18	0.50
51	0.00	0.00	0.00	0.00	0.00	0.00	0.00	0.00	0.00	0.00	0.00	0.00	0.00	0.00	0.00	0.00	0.00	0.00	0.00	0.00	0.00	0.00	0.00	0.00	0.00	0.00	0.00	0.00	0.00
52	0.50	0.00	0.00	0.00	0.00	0.00	0.00	0.50	0.00	0.00	0.00	0.33	0.33	0.00	0.00	0.00	0.00	0.00	0.00	0.00	0.00	0.00	0.00	0.00	0.00	0.00	0.00	0.00	0.00
53	0.00	0.00	0.00	0.00	0.00	0.00	0.00	0.00	0.00	0.00	0.00	0.00	0.00	0.00	0.00	0.00	0.00	0.00	0.00	0.00	0.00	0.00	0.00	0.00	0.00	0.00	0.00	0.00	0.00
54	0.00	0.00	0.00	0.00	0.00	0.00	0.00	0.00	0.50	0.33	0.00	0.00	0.00	0.00	0.00	0.00	0.00	0.00	0.00	0.00	0.00	0.00	0.00	0.00	0.00	0.00	0.00	0.00	0.50
55	0.00	0.00	0.00	0.00	0.00	0.00	0.00	0.00	0.00	0.00	0.00	0.00	0.00	0.00	0.00	0.00	0.00	0.00	0.00	0.00	0.00	0.00	0.00	0.00	0.00	0.00	0.00	0.00	0.00
56	0.33	0.00	0.00	0.00	0.00	0.00	0.00	0.00	0.00	0.00	0.00	0.00	0.00	0.00	0.00	0.00	0.00	0.00	0.00	0.00	0.00	0.00	0.00	0.00	0.00	0.00	0.00	0.00	0.00
57	0.00	0.00	0.00	0.00	0.00	0.00	0.00	0.00	0.00	0.00	0.00	0.00	0.00	0.00	0.00	0.00	0.00	0.00	0.00	0.00	0.00	0.00	0.00	0.00	0.00	0.00	0.00	0.00	0.00
58	0.00	0.00	0.00	0.00	0.00	0.00	0.00	0.00	0.19	0.00	0.00	0.00	0.00	0.00	0.00	0.00	0.00	0.00	0.00	0.00	0.00	0.00	0.00	0.00	0.00	0.00	0.00	0.00	0.00
59	0.00	0.00	0.00	0.00	0.00	0.00	0.00	0.00	0.00	0.00	0.00	0.00	0.00	0.00	0.17	0.00	0.00	0.00	0.00	0.00	0.00	0.00	0.00	0.00	0.00	0.00	0.00	0.00	0.00
60	0.00	0.00	0.00	0.00	0.00	0.00	0.00	0.00	0.00	0.00	0.00	0.00	0.00	0.50	0.33	0.00	0.00	0.25	0.50	0.00	0.25	0.25	0.25	0.25	0.50	0.00	0.00	0.00	0.00
61	0.00	0.00	0.00	0.00	0.00	0.00	0.00	0.00	0.00	0.00	0.00	0.00	0.00	0.00	0.00	0.00	0.00	0.00	0.50	0.00	0.00	0.00	0.00	0.00	0.00	0.00	0.00	0.00	0.00
62	0.33	0.00	0.00	0.00	0.00	0.00	0.00	0.00	0.00	0.00	0.00	0.00	0.00	0.00	0.00	0.00	0.00	0.00	0.00	0.00	0.00	0.00	0.00	0.00	0.00	0.00	0.00	0.00	0.00
63	0.17	0.00	0.00	0.00	0.00	0.00	0.00	0.00	0.25	0.00	0.00	0.00	0.00	0.00	0.00	0.00	0.00	0.00	0.00	0.00	0.00	0.00	0.00	0.00	0.00	0.00	0.00	0.00	0.00
64	0.33	0.00	0.00	0.00	0.00	0.00	0.00	0.00	0.00	0.00	0.00	0.00	0.00	0.00	0.00	0.00	0.00	0.00	0.00	0.00	0.00	0.00	0.00	0.00	0.00	0.00	0.00	0.00	0.00
65	0.00	0.00	0.00	0.00	0.00	0.00	0.00	0.00	0.00	0.00	0.00	0.00	0.00	0.00	0.00	0.00	0.00	0.00	0.00	0.00	0.00	0.00	0.00	0.00	0.00	0.00	0.00	0.00	0.00
66	0.00	0.00	0.00	0.00	0.00	0.00	0.00	0.00	0.00	0.00	0.00	0.00	0.00	0.00	0.00	0.00	0.00	0.00	0.00	0.00	0.00	0.00	0.00	0.00	0.00	0.00	0.00	0.00	0.00
67	0.33	0.00	0.00	0.00	0.00	0.00	0.00	0.00	0.00	0.38	0.50	0.00	0.25	0.00	0.00	0.00	0.00	0.00	0.00	0.00	0.00	0.00	0.00	0.00	0.00	0.00	0.00	0.00	0.00
68	0.00	0.00	0.00	0.00	0.00	0.00	0.00	0.00	0.00	0.00	0.00	0.00	0.00	0.00	0.17	0.00	0.17	0.17	0.17	0.00	0.00	0.00	0.00	0.00	0.00	0.00	0.17	0.00	0.00
69	0.00	0.00	0.00	0.00	0.00	0.00	0.00	0.00	0.00	0.00	0.00	0.00	0.00	0.00	0.17	0.00	0.00	0.13	0.00	0.00	0.00	0.00	0.00	0.11	0.00	0.00	0.17	0.11	0.00
70	0.00	0.00	0.00	0.00	0.00	0.00	0.11	0.00	0.00	0.00	0.00	0.00	0.00	0.00	0.17	0.00	0.00	0.13	0.00	0.00	0.00	0.00	0.00	0.00	0.00	0.00	0.17	0.00	0.00
71	0.00	0.00	0.00	0.00	0.00	0.00	0.00	0.00	0.00	0.00	0.00	0.00	0.00	0.00	0.00	0.00	0.00	0.00	0.00	0.00	0.00	0.00	0.00	0.00	0.00	0.00	0.00	0.00	0.00
72	0.33	0.00	0.00	0.00	0.00	0.00	0.00	0.00	0.00	0.00	0.00	0.00	0.00	0.00	0.00	0.00	0.00	0.00	0.00	0.00	0.00	0.00	0.00	0.00	0.00	0.00	0.00	0.00	0.00
73	0.33	0.00	0.00	0.00	0.00	0.00	0.00	0.00	0.00	0.25	0.50	0.00	0.00	0.00	0.00	0.00	0.00	0.00	0.00	0.00	0.00	0.00	0.00	0.00	0.00	0.00	0.00	0.00	0.00
74	0.50	0.00	0.00	0.00	0.00	0.00	0.00	0.00	0.00	0.00	0.50	0.00	0.00	0.00	0.00	0.00	0.00	0.00	0.00	0.00	0.00	0.00	0.00	0.00	0.00	0.00	0.00	0.00	0.00
75	0.33	0.00	0.00	0.00	0.00	0.00	0.00	0.00	0.00	0.25	0.50	0.25	0.25	0.00	0.00	0.00	0.00	0.00	0.00	0.00	0.00	0.00	0.00	0.00	0.00	0.00	0.00	0.00	0.00

RAND *MG979-A.8*

APPENDIX B

Estimation of Marginal Implementation Cost for Systems Derivable from ATOs

This surrogate Delphi exercise was conducted by two study team members. Both were closely involved in the cost estimation for TAS-1. Although the implementation cost includes many components for demonstrating, acquiring, fielding, operating, and maintaining the systems derived from an ATO, the key drivers are the unit cost of a system and the number of units deployed over an assumed planning horizon of 20 years in order to meet the FOC requirements. Both TAS-1 and the present study follow a methodology based on the concept of marginal cost over the legacy system that the ATO system replaces. The two evaluators also have a good notion of the level of the marginal implementation cost based on two key factors: (1) the unit cost over that of a legacy system[1] and (2) the number of units deployed.

The two evaluators conducted the first round independently. After they reviewed each other's first-round evaluation and considered their differences in estimations but without any discussion between them, they redid their evaluations as their second round. Then, whenever there remained a difference in evaluating a system, they discussed why they differed. In all cases, they came to a consensus.[2] The results are shown in Table B.1.

For a real Delphi exercise, we expect five to ten evaluators and four rounds. It is preferable for the evaluators to be familiar with the ATOs and the marginal implementation costs of their systems. If this is not the case, the organizer can select examples from TAS-1 that illustrate the typical unit costs of various classes of systems and the numbers of units deployed to meet the FOC capability requirements.

---

[1] If the ATO system provides a new capability on a new, as opposed to legacy, platform, the marginal unit cost of the ATO system is the same as its unit cost (i.e., the legacy system cost is zero). Moreover, if an ATO aims to save legacy system costs, the marginal implementation cost will be negative. Allowing negative cost is an important feature of our approach, because we believe that it is the best way to properly capture the contribution of a cost-saving ATO.

[2] In a real Delphi exercise, if the evaluators do not come to a consensus on some of the ATOs, planners can take the average and use the deviation to examine sensitivity of findings due to variation in these estimates.

Table B.1
Marginal Implementation Cost Expressed in Grades for ATOs 1–75

ATO Number	ATO Name	Initial Estimates (Evaluator One)	Initial Estimates (Evaluator Two)	Revised Estimates (Evaluator One)	Revised Estimates (Evaluator Two)	Final Consensus
1	Network electronic warfare ATO	0	3	3	3	3
2	Mine and IED detection ATO	1	3	3	3	3
3	Rotorcraft survivability ATO	3	2	3	3	3
4	Kinetic Energy Active Protection System ATO	3	3	3	3	3
5	Passive Infrared Cueing System ATO	3	4	3	4	4
6	Extended-area protection and survivability ATO	0	2	2	2	2
7	Dissemination of advanced obscurants ATO	2	2	2	2	2
8	Pulse power for the FCS ATO	4	2	2	2	2
9	Vehicle armor technology ATO	3	4	3	4	3
10	Solid-state laser technology ATO	4	4	4	4	4
11	Countermine and IED neutralization ATO	2	4	2	4	4
12	Vision protection ATO	3	2	3	2	2
13	Modular protective systems for Future Force assets ATO	3	3	3	3	3
14	Wide area airborne minefield detection ATO	0	4	4	4	4
15	Third-generation IR technologies ATO	4	4	4	4	4
16	Distributed Aperture System ATO	4	4	4	4	4
17	Suite of sense-through-the-wall systems ATO	3	3	3	3	3
18	Multimission radar ATO	4	4	4	4	4
19	All-Terrain Radar for Tactical Exploitation of Moving Target Indicator and Imaging Surveillance System ATO	3	4	3	4	3
20	Distributed imaging radar technology for continuous battlefield imagery ATO	3	3	3	3	3
21	Class-II UAV EO payloads ATO	2	4	2	4	4

Table B.1—Continued

ATO Number	ATO Name	Initial Estimates (Evaluator One)	Initial Estimates (Evaluator Two)	Revised Estimates (Evaluator One)	Revised Estimates (Evaluator Two)	Final Consensus
22	Objective pilotage for utility and lift ATO	3	4	3	3	3
23	Soft-target exploitation and fusion ATO	1	0	1	1	1
24	Low-cost, high-resolution IR focal plane arrays ATO	–1	–1	–1	–1	–1
25	Human infrastructure detection and exploitation ATO	4	4	4	4	4
26	Tactical wireless network assurance ATO	1	1	1	1	1
27	Networked enabled command and control ATO	1	2	1	1	1
28	Tactical mobile networks ATO	1	3	1	1	1
29	Tactical network and communications antennas ATO	1	2	1	1	1
30	Battlespace terrain reasoning and awareness—battle command ATO	1	2	1	1	1
31	NLOS-LS launch system technology ATO	4	2	2	2	2
32	Mounted Combat System and Abrams Ammunition System technologies ATO	3	3	3	3	3
33	Common smart submunition ATO	3	0	0	0	0
34	Nonlethal payloads for personnel suppression ATO	1	4	3	4	3
35	Electromagnetic gun technology maturation and demonstration ATO	3	4	3	4	3
36	Fuze and power for advanced munitions ATO	2	2	2	2	2
37	Microelectromechanical systems inertial measurement unit ATO	–2	–2	–2	–2	–2
38	Smaller, lighter, cheaper munitions components ATO	–1	–1	–1	–1	–1
39	Hardened combined effects Penetrator warheads ATO	1	1	1	1	1
40	Insensitive munitions technology ATO	1	1	1	1	1

Table B.1—Continued

ATO Number	ATO Name	Initial Estimates (Evaluator One)	Initial Estimates (Evaluator Two)	Revised Estimates (Evaluator One)	Revised Estimates (Evaluator Two)	Final Consensus
41	Novel energetic materials for the Future Force ATO	1	2	1	2	2
42	Missile propulsion technology ATO	2	2	2	2	2
43	Missile seeker technology ATO	2	2	2	2	2
44	Automated Critical Care Life Support System ATO	2	2	2	2	2
45	Fluid resuscitation technology to reduce injury and loss of life on the battlefield ATO	2	2	2	2	2
46	Vaccines and drugs to prevent and treat malaria ATO	1	0	1	1	1
47	Vaccines to prevent diarrhea ATO	1	0	1	1	1
48	Vaccine for the prevention of military HIV infection ATO	1	2	1	1	1
49	Biomedical enablers of operational health and performance ATO	1	1	1	1	1
50	Robotics collaboration ATO and near autonomous unmanned systems ATO	2	2	2	2	2
51	UAV System technologies ATO	3	3	3	3	3
52	Army/DARPA enabling technologies for the FCS ATO	4	4	4	4	4
53	Manned/unmanned rotorcraft enhanced survivability ATO	0	0	0	0	0
54	Future Force Warrior ATD	2	2	2	2	2
55	Soldier mobility vision systems ATO	2	2	2	2	2
56	Nutritionally optimized first strike ration ATO	0	0	0	0	0
57	Soldier protection technologies ATO	3	3	3	3	3
58	Mounted/dismounted soldier power ATO	4	4	4	4	4
59	Infantry warrior simulation ATO	0	0	0	0	0
60	Leader adaptability ATO	0	0	0	0	0
61	Strategies to enhance retention ATO	–1	–2	–1	–1	–1

Table B.1—Continued

ATO Number	ATO Name	Initial Estimates (Evaluator One)	Initial Estimates (Evaluator Two)	Revised Estimates (Evaluator One)	Revised Estimates (Evaluator Two)	Final Consensus
62	Precision airdrop–medium ATO	2	2	2	2	2
63	Hybrid electric for the FCS ATO	2	2	2	2	2
64	Advanced lightweight track ATO	2	2	2	2	2
65	Joint rapid airfield construction ATO	1	0	1	1	1
66	JP-8 reformer for alternate fuel sources ATO	4	4	4	4	4
67	Prognostics and diagnostics for operational readiness and condition-based maintenance ATO	3	4	3	3	3
68	Learning with adaptive simulation and training ATO	0	0	0	0	0
69	Scaleable embedded training and mission rehearsal ATO	0	0	0	0	0
70	Severe trauma simulation ATO	1	0	1	1	1
71	Joint enabled theater access–sea ports of debarkation ACTD	2	2	2	2	2
72	Tactical wheeled vehicle fleet modernization and future tactical truck systems ACTD	2	2	2	2	2
73	Adaptive joint C4ISR node ACTD	–1	–1	–1	–1	–1
74	Theater effects–based operations ACTD	2	2	2	2	2
75	Joint Modular Intermodal Distribution System JCTD	2	2	2	2	2

Moreover, cost estimates from the Delphi exercise and other means should be provided to the program managers of relevant ATOs so that they can comment on the accuracy of these estimates. TAS-1 suggests an iterative procedure for estimating lifecycle costs and ATO contributions to capability requirements during the Army S&T community's annual technology planning, review, and oversight process. This iterative procedure can improve the accuracy of these estimates. Most important, the procedure would make the community more comfortable in incorporating these estimates into their S&T planning and decisions. The alternative of ignoring lifecycle costs in ATOs, the highest-priority S&T programs, is no longer compatible with the intent of the

Defense Acquisition System and the desire of Army senior management (DoD, 2003; U.S. Army, 2006, pp. 26–27).

Bibliography

Brown, Bradford, *Operation of the Defense Acquisition System, Statutory and Regulatory Changes*, Kettering, Ohio: Defense Acquisition University, December 8, 2008.

Chow, Brian G., Richard Silberglitt, and Scott Hiromoto, *Toward Affordable Systems: Portfolio Analysis and Management for Army Science and Technology Programs*, Santa Monica, Calif.: RAND Corporation, MG-761-A, 2009. As of October 21, 2010:
http://www.rand.org/pubs/monographs/MG761/

DoD—*see* U.S. Department of Defense.

England, Gordon, "Capability Portfolio Management Test Case Roles, Responsibilities, Authorities, and Approaches," Washington, D.C.: U.S. Department of Defense, September 14, 2006.

Office of the Under Secretary of Defense, Comptroller, *DoD Financial Management Regulation, Volume 2B: Budget Formulation and Presentation*, "Research, Development and Evaluation Appropriations," Chapter 5 (July 2008), July 6, 2000. As of October, 27, 2010:
http://www.dod.mil/comptroller/fmr/02b/

Public Law 111-23, Weapon Systems Acquisition Reform Act of 2009, "Title II, Acquisition Policy," May 22, 2009.

Silberglitt, Richard, and Lance Sherry, *A Decision Framework for Prioritizing Industrial Materials Research and Development*, Santa Monica, Calif.: RAND Corporation, MR-1558-NREL, 2002. As of October 21, 2010:
http://www.rand.org/pubs/monograph_reports/MR1558/

Silberglitt, Richard, Lance Sherry, Carolyn Wong, Michael S. Tseng, Emile Ettedgui, Aaron Watts, and Geoffrey Stothard, *Portfolio Analysis and Management for Naval Research and Development*, Santa Monica, Calif.: RAND Corporation, MG-271-NAVY, 2004. As of October 21, 2010:
http://www.rand.org/pubs/monographs/MG271/

TAS-1—*see* Chow, Silberglitt, and Hiromoto, 2009.

TAS-2—*This is the current study.*

U.S. Army, Assistant Secretary of the Army, Financial Management & Comptroller, *Cost and Performance Portal (CPP)*, "Capability-Based Cost Analysis Tools: Capabilities Knowledge Base (CKB)," data files, undated. Accessed May 7, 2010.

———, *Military Operations: Force Operating Capabilities*, Fort Monroe, Va.: Training and Doctrine Command, pamphlet 525-66, July 1, 2005. As of October 21, 2010:
http://www.tradoc.army.mil/tpubs/pams/p525-66.pdf

———, *2006 Army Modernization Plan: Building, Equipping, and Supporting the Modular Force*, 2006. As of October 22, 2010:
http://www.army.mil/features/MODPlan/2006/

———, *Army Science & Technology Master Plan—Executive Summary*, Office of the Deputy Assistant Secretary of the Army for Research and Technology, Fort Belvoir, Va.: Defense Technical Information Center, 2007, p. I-8. As of October 21, 2010:
http://www.carlisle.army.mil/dime/documents/JPLD_AY08_Lsn%207_Reading%204_ASTMP.pdf

U.S. Department of Defense (DoD), *Acquisition: Army Transition of Advanced Technology Programs to Military Applications*, Arlington, Va.: Office of the Inspector General, D-2002-107, June 14, 2002. As of October 27, 2010:
http://www.dodig.mil/audit/reports/fy02/02-107.pdf

———, Under Secretary of Defense for Acquisition, Technology and Logistics (USD [AT&L]), *Operation of the Defense Acquisition System*, Department of Defense Instruction 5000.2, May 12, 2003. As of October 21, 2010:
http://www.carlisle.army.mil/dime/documents/JPLD_AY08_Lsn%207_Reading%203_DoDI%20 5000-2.pdf